TURN SIGNALS ARE THE FACIAL EXPRESSIONS OF AUTOMOBILES

TURN SIGNALS ARE THE FACIAL EXPRESSIONS OF AUTOMOBILES

Donald A. Norman

A William Patrick Book

Addison-Wesley Publishing Company

Reading, Massachusetts   Menlo Park, California   New York
Don Mills, Ontario   Wokingham, England   Amsterdam   Bonn
Sydney   Singapore   Tokyo   Madrid   San Juan
Paris   Seoul   Milan   Mexico City   Taipei

"This Is Just to Say" is taken from *The Collected Poems of William Carlos Williams, 1909–1939, vol. 1.* Copyright 1938 by New Directions Publishing Corporation. Reprinted by permission of New Directions Publishing Corporation.

Library of Congress Cataloging-in-Publication Data

Norman, Donald A.
    Turn signals are the facial expressions of automobiles / Donald A. Norman.
        p.   cm.
    "A William Patrick book."
    Includes bibliographical references and index.
    ISBN 0-201-58124-8
    ISBN 0-201-62236-X (pbk.)
    1. Technology.   I. Title.
T47.N67   1992
600—dc20   91-38762
                                                                                CIP

Cover design by Stephen Gleason
Set in 10½-point Trump Medieval by Shepard Poorman Communications

1 2 3 4 5 6 7 8 9-MA-96959493
First printing, April 1992
First paperback printing, February 1993

To my parents:

Miriam F. Norman
Noah N. Norman

Contents

Preface

SOME people watch birds. Others watch people. Some watch cars or boats, or sports. I watch technology, especially the small, common, everyday variety. In particular, I watch the way people interact with technology. I am not happy with what I see. Much of modern technology seems to exist solely for its own sake, oblivious to the needs and concerns of the people around it, people who, after all, are supposed to be the reason for its existence.

The dehumanizing nature of modern technology has long been a theme for concern. My goal is neither to attack nor to defend but to understand just how the interaction between person and technology takes place, to discover where and why difficulties appear, and then to try to do something about them. You might say that my goal is to socialize technology, to humanize technology.

Society has reached the point where it would be completely unable to function without technology. Our social judgments, our skills, and even our thoughts are indelibly affected by the nature of the technology that supports us. Worse, the impact is so pervasive, so subtle, that we are often unaware of how many of our beliefs have been affected by the arbitrary nature of technology.

This book touches upon some of the critical issues. The chapters are meant to inform as they amuse. Some are deliberately provocative. But I am serious about the main message: Technology tends to dehumanize. This is not a necessary part of technology, but it relentlessly encroaches upon us unless we exert caution. Technologists tend to create what technology makes possible without full regard for the impact on human society. Moreover, technologists are experts at the mechanics of their technology but often are ignorant of and sometimes even disinterested in social concerns.

I believe that we can create a humane technology, one that serves and enhances people and society. But this won't happen unless educated, concerned people take a hand in the design and application of our technology. This book provides some starting points. It provokes, scolds, and speculates, all with the aim of increasing our sensitivity to the dehumanizing aspects of technology, and helping us to realize that we can cause it to be otherwise.

Acknowledgments

I AM grateful to many for their assistance and comments on these essays. My academic colleagues have provided much of my initial stimulation and my interactions with industrial developers and researchers have increased my appreciation for the difficulties of bringing technologies to the marketplace. I am grateful to all my colleagues.

I especially thank Mike Shafto and Ev Palmer at NASA Ames Research Center (who also support my research through the NASA Aviation Safety Program), my colleagues at Apple Computer, Inc., and Digital Equipment Corporation, who have also helped support this research, and especially to Tom Erickson and Jim Spohrer at Apple for their comments on the manuscript. Julie Norman was a perceptive and properly critical reader of the early drafts. William Patrick at Addison-Wesley provided a useful scolding about the excesses of earlier drafts, telling me what was nonsense, helping me shape these essays into coherence, yet without losing either the fun or the message. This book might not have been written at all if it were not for the encouragement of Sandra Dijkstra, my literary agent, who was always there when needed.

At the University of California, San Diego, Michael Cole continues to educate me about the nature of mind and culture; Edwin Hutchins has provided critical insights about the way that cognition is distributed over time, space, people, and things; and Hank Strub, Emmy Goldknopf, and Cyndi Norman kept plying me with critical readings, critiques, and queries, all to the eventual betterment of my own understanding. Comments have come from all over the world through the power of the electronic mail and bulletin boards that link the universities of the world (see, for example, Chapter 4, which is entirely the result of electronic mail interaction).

A book is the joint effort of many people. My experience has provided a wonderful example of how the interaction of cognition, technology, and social interactions can merge to aid human activity. Technology can be effective, supportive, and humanizing, if only we insist that it be so.

D.A.N.

Del Mar, California

Chapter Notes and Book Design

NOTHING seems to create more controversy about the design of a book than the placement of notes for each chapter. Academic readers are used to seeing notes at the bottom of relevant pages as footnotes. Trade publishers do not approve. They feel that notes distract, breaking the flow of reading. They prefer to hide the notes at the end of the book out of sight, but still available for the serious reader.

Many of my readers have complained vociferously. The notes are hard to find, they say, and it is particularly disruptive to keep two place markers, one for text, one for notes. Alas, I know of no data relevant to the relative proportion of readers who might fall into each category, the one described by the publishers or the one corresponding to those who write me. Each side argues that the other constitutes a tiny minority.

For this book, we are trying a compromise. Important comments have been put directly in the text. Notes are used primarily for references and acknowledgments. I have kept substantive comments to a minimum. This should minimize the need to turn to a note. Each note specifies the chapter number, name, and page from which it has come, with a short quote from the text indicating the section to which it refers. This makes it easy to go from the notes back to the text, a feature many readers requested.

There still remains the question of how to tell the reader that there is a relevant note at the back of the book. Traditional footnote symbols are thought to be disruptive, diverting the reader's attention to the note, usually to find that it does not contain essential information. We are trying a more subtle signal. The mark "*" appears whenever there is a substantive comment in the chapter notes. Notes that contain only a citation to the published literature or an acknowl-edgment are not marked in the text.

TURN SIGNALS ARE THE FACIAL EXPRESSIONS OF AUTOMOBILES

1

I Go to a Sixth Grade Play

I WENT to a sixth grade play. It was a small play, at a small school. Only the sixth grade was involved, so we were in a relatively small auditorium, with approximately fifty folding chairs crammed together on the floor. If there had been only fifty parents present, it would have been crowded. But in addition to the parents, we had the video cameras.

Some parents came with camera alone but then had to seek an electric outlet to provide power. Some parents had cameras that worked on batteries—fully charged batteries, they smugly declared— and they had no need for electric outlets. Some brought tripods to steady their cameras and then had to squeeze themselves and others in order to set them up. One set of parents came fully equipped with carrying case, tripod, lights, and a long, bright orange extension cord, neatly rolled up on a special carrying reel. Microphones. Tripods. Cameras. Lights. Lights? To bathe the audience or the play?

Whenever I travel, I watch with awe and amazement people overloaded with recording devices. It used to be fun to talk of national stereotypes. Traveling Americans with cameras always at their sides, traveling Germans with cameras at their sides, traveling Japanese with cameras always at their eyes. But the fever recognized no national boundaries, and the early stereotypes soon dissolved into near univer- sal behavior. Still cameras, small pocket-sized automatic cameras, each with built-in, automatic flash. Flashing lights popping off here, pop- ping off there. Flashing for people taking photographs of dramatic sunsets.

In the old days those of us who were technically sophisticated used to scoff at those who used a flash to illuminate events far in the distance, let alone a flash to illuminate the setting sun. Today the sneers are gone, for the cameras have taken over, and even I—technically literate in film speeds and gamma, f-ratios and exposure indices—find my little automatic camera desperately engaging its electronic flash, in an effort to illuminate the sky that is not only futile but counterproductive at that, for if it succeeds, it will ruin the photograph.

Ah yes, once upon a time there was an age in which people went to enjoy themselves, unencumbered by technology, with the memory of the event retained within their own heads. Today we use our artifacts to record the event, and the act of recording then becomes the event. Days later we review the event, peering at the tape, film, or video in order to see what we would have seen had we been looking. We then show the recorded event to others so they too can experience what we would have seen had we been looking. Even if they don't care to experience it, thank you.

Vicarious Experiencing

The technologies for recording events lead to a curious result. Vicarious experience, I call it. Vicarious experience, even for those who were there. In this context "vicarious" means to experience an event through the eyes (or the recording device) of another. Yet here we have the real experiencer and the vicarious experiencer being the same person, except that the real experiencer didn't have the original experience because of all the activity involved in recording it for the later, vicarious experience. If you get what I mean. Of course, a vicarious experience is never the same as a real one. We are so busy manipulating, pointing, adjusting, framing, balancing, and preparing that the event disappears. The artifact becomes the event.

The flaw becomes apparent when the technology fails, when we miss both the original experience (because it was transformed into the experience of manipulating the technology) and the playback, when the technological failure leaves us with nothing to play back.

In March 1970 the Amateur Astronomers Association of New York conducted an expedition to Virginia Beach, Virginia, to view a solar eclipse. A participant later wrote in the *Eyepiece*, the club newspaper:

I . . . spent almost the entire time during totality, attempting to load a camera in the dark. My wife was standing near me. She was full of excitement, yelling all sorts of incredibly beautiful things that were supposedly happening up there. At one point, she just yanked me and said, 'Look, look there's Venus. THERE IS VENUS!' and sure enough, when I looked up, there was Venus. But as I looked up I lost the grip on the film spool, which was half-laced in the sprocket. The thing fell out and rolled off somewhere and I spent practically the rest of totality (of the eclipse) groping for it in the dark.

I, therefore, came to the conclusion, and it represents a bit of wisdom, gained at great expense and hardship, which I feel compelled to pass on to other enthusiasts, and that is, the best possible way to see an eclipse is simply to LOOK AT IT!

I used to travel with a camera myself, always careful to record all significant events, spending what to my fellow travelers must have seemed like hours, finding just the correct vantage point, the correct lighting to capture the moment. I ended up with hundreds of photographic slides, maybe thousands. So many that I no longer had time to organize them, to arrange them—no time even to look at them. I discovered that I was spending much of my traveling time recording events for a future that never happened. Perhaps it was that magnificent photograph in Morocco that convinced me. I got it all perfect: two camels walking by, palm trees in the background, the sun setting. "I'll make a large print and frame it: This will be our memory of Morocco," I told my wife.

Today I'm not even certain that I ever looked at the resulting photograph. Happily, the moment is etched forever in my memory, perhaps in better form than the photograph (which for all I know has telegraph wires and everything out of focus).

Probably we've all seen a wedding reception, an event meant to be full of spontaneous expressions of joy, transformed by the photographer into a series of staged events. "Kiss the bride." "Again, please." "Cut a piece of the wedding cake." "Would the bride feed the groom?" "Move out of the way of the camera." It is amazing how tolerant we have become of this manipulation of experience: The act of recording taking precedence over the event.

Technology's Short Life Span: Are Those Memories Really Forever?

Suppose that we deliberately seek to record critical events in our lives so that in future years we might revisit these important times. I record the appearances of my family: weddings and births, school events and graduations, trials and tribulations. Yes, human memory has wonderful powers, but it has (at least) two major limitations:

The first is that our memories of people are always being continually revised, continually adjusted. We remember people as they are today, not as they were in the past. It is hard to remember what a close companion looked like twenty years ago: The image of today overwhelms the image of yesterday.

This is actually a good thing, for otherwise we would not recognize people when they changed clothes, or hairstyle, or even posture. I have had a beard for scores of years, so long that neither I nor my wife remembers what I looked like without one. I have to resort to photographs to remember my own face (and my children cannot believe these are photographs of me).

On the other hand, if someone grows a beard and keeps it for a year or so, then one day suddenly shaves it off, the surprising thing is how few people who knew him before and during the beard notice its absence. Most people notice something—"Did you change your glasses?"—but not the absence of the beard. Memory is extremely flexible, always modifying its image. Even the events that we remember with great clarity may be deceptive, for what we take to be a complete and accurate memory may differ substantially from the actual happening. Our memories reveal their flexibility with everyday experiences. Here, memory can be amazingly insensitive to change, adjusting itself to the normal changes in appearances. This adjustment process can also make it difficult to recover how the people we interact with every day appeared in the past.

We are slow to notice changes in our daily acquaintances. I live with my children every day for years and hardly notice that they are

4

growing taller and more mature at a rapid pace. It is only when I go away for a lengthy trip and return a week or more later that I am astonished by how tall and grown-up they have become. Families often do not notice the deteriorating behavior and activities of their members, so that major symptoms of diseases can go unnoticed, even though they are immediately obvious to a newcomer. Presence breeds familiarity, encouraging the perception that how it is now is how it has always been.

The second limitation of human memory is that it is a private experience. Whether our memory of an event is accurate or not, vivid or faint, we cannot share it with others exactly as we experienced it.

And so for these and other reasons, we turn to technology to provide a long-lasting record of our experiences and moments in a way that is unchangeable over time as well as accessible to others. But how long-lasting is the record of technology?

In my studies of technology I have distinguished between two modes of representing information. An artifact can have a surface representation or an internal one. A *surface* representation is like the words on the page of this book. The information is all there on the surface: What you see is what there is. Photographs, drawings, letters, and books all exploit surface representations.

The advantage of surface representation is that no technology is needed to experience it: All we have to do is look, and there it is. All that matters is that the artifact still exists and that the marks on the surface are still visible.

This is not as simple as it may seem. In the past hundred years, most paper has been manufactured using an acid-based technology that destroys the paper itself after some time. Most of us have experienced the deterioration of newspapers: Leave one exposed to the light, and a year or two later it crumbles and fades away. Books and personal notes have a similar, if slower, deterioration. Today paper manufacturers recognize this problem and most quality books are published on acid-free papers. But notepads are not.

Photographs also fade. Slides, color prints, and black-and-white photographs last only as long as their chemically activated marks stay there, as long as the paper on which they are printed is strong. Prints last longer if you don't expose them to the light. So photographs last longer if you don't look at them. Fortunately, we can expect books, notes, and photographs to last approximately 100 years, long enough for most purposes, certainly longer than the life span of the individuals who made them.

The story is quite different if we entrust our records to *internal* artifacts, artifacts where the information is stored invisibly inside the material, where technology is required to retrieve it.

Remember the days when we recorded voices on spools of wire, rapidly passing before a magnetic head? You probably don't remember the forerunner of today's tape recorder. But that is just the point. If I recorded my baby's first words on a wire recorder, or perhaps had the sole surviving recording of President Roosevelt's farewell words to Joseph Stalin and Winston Churchill, they would be virtually unrecoverable. Unless someone could discover a working version of the wire recorder (or could make one), there would be no way to recover those sounds again. Or suppose I had recorded the sounds on one of Edison's wax cylinder phonograph records.

Do you think the videotape technology we use today will be around fifty years from now? No way. Today's videotapes will not be playable. Nor will our audiotapes. Or our phonograph records. The same, sad story is true of all technologies that use internal storage—that is, any recording device that requires another technology to reproduce it.

The same problems have already affected science. In the United States, the National Aeronautics and Space Administration (NASA) has tens of thousands of reels of magnetic tape, carefully preserved, with data recorded from space and satellites. From those data we could follow in detail the earth's weather patterns and cycles of plant growth, or water, or pollution. Except that they are recorded on an old-fashioned tape with an old-fashioned format that can no longer be read by today's computers.

I have not even discussed the fact that the magnetic particles on computer, video, and audiotapes and wire recordings do not last forever. They slowly migrate, losing their signals. And the tape itself cracks and deteriorates.

Nothing in this world is permanent; nothing lasts forever. But most of us had not counted on technology's assistance in hastening the demise of our memories. But there is a positive side to the use of recording devices: situations where the device intensifies the experience. Most of the time this takes place only with less sophisticated artifacts: the sketch pad, the painter's canvas, and the notebook. Those who benefit from these intensifying artifacts are usually artists and writers. But the benefits are not restricted to these people, and with the proper frame of mind, can extend to photographers as well. The critical point is that with these artifacts, the act of recording forces us to look and experience with more intensity and enjoyment than might otherwise be the case.

I was first exposed to the contrast between the obstructionist and the intensifying artifact on a journey to the Yellow Mountain (Huang Shan) in the province of Anhui in China. This was a long trek, one that for many Chinese was a lifetime dream. For my family and our hosts, it was a five-day trip: one long day's drive in a crowded Toyota van, three days of hiking, then the day-long drive back to the University of Science and Technology of China in Hefei. On the mountain itself, the long trail was crowded, with an endless line of Chinese climbing patiently upward. The only other Westerner we encountered turned out to be an American photographer for the National Geographic Society.

The peaks were amazing sights of beauty. The clouds formed a sea, through which small peaks here and there poked their heads, as if they were islands. In fact, we spent our first night on the mountains at a location called "The North Sea," the name referring to the way in which the mountaintops look like islands within the sea of clouds. The rocks—and for that matter, the vegetation—were distorted in what I can only describe as a Chinese manner. I suddenly realized that the grotesque shapes and forms in many Chinese scroll paintings were not from the imagination: That is how the countryside really looks.

How to record such a wondrous event? My family and I, true to our tradition of mental recordings, simply walked around in awe and wonder. The modern Chinese all had cameras, and they photographed one another standing against the traditional monuments of the Yellow Mountain. Actually, we turned out to be one of the main sights, especially my (then) five-year-old, blond-haired, blue-eyed son. The Chinese crowded around to have their pictures taken with this strange and foreign person (pinching and tugging at his hair to make sure it was real). The people crowding around us and Eric remain part of our memory of that trip. As for our fellow travelers with cameras, I suspect their pictures were not intended to remember the experience as much as to prove the experience: their proof that they made this journey. Look, here I am in front of Purple Cloud Peak; here I am with the foreign yellow-haired Western boy. See, it must be true.

But what impressed me more was the large crowd of artists. Early in the morning, as the sun's rays broke through the mist and the top of the mountains peeked through the sea of clouds, the artists scrambled to vantage points where they would sit patiently for hours—drawing, painting, recording.

Why would anyone want to draw when they could photograph and get an exact image? Because the drawing made the experience personal. And the act of drawing requires a degree of concentration and study that intensifies the experience. So even if the drawing is thrown away and never looked at, the active participation in its creation makes the experience of the Yellow Mountain more intense, more personal, more enjoyable.

Unlike the wedding photographers I described earlier, some people who use cameras can also have the intensifying experience. These are the photographers who are artists, the ones who examine and study every scene with care, intensely noting this feature and that, using their artistic sense to plan each photograph. For these photographers, the experience is enhanced in much the same way that it is for the person sketching the scene: intense concentration, a feeling of participating in the event. Drawings, paintings, and photographs made from these perspectives can all reveal aspects of the scene not apparent even to others who were there. It is not the artifact that is obstructionist; it is the way in which it is applied.

There is a negative aspect to this enhanced perception. I would like to be able to tell you how the active recording of the event, whether in picture or words, makes the observer ever more sensitive to the critical details and experience and better able to describe it. Alas, although the statement is true, the flaw is that the act of recording is often slower than the event itself, so the act of concentrating upon one aspect of one event guarantees that all other aspects will be missed. At the Yellow Mountain, changes occurred slowly, but even so, an hour spent drawing the southerly view would prevent an artist from experiencing the other views, each being different, each continually revealing nuances and subtleties of the experience.

In fact, my note-taking for this essay makes the point. The notes that became this essay were triggered by my examination of the video cameras at my son's sixth grade play. I watched with great interest the goings-on of the parents. I took out my pocket notebook and started to write my thoughts. But midway through the note-taking, on the sixth page, I was interrupted by a poke in the ribs. My wife pointed out that my rapt attention to the video cameras and my notepad made me ignore our son, who was at that very moment on the stage reciting his part in the play.

But certainly we will do better in the future, will we not? Notepads will actually be computers, accepting handwriting or short-hand, keyboard symbols or speech. These technologies will change the act of note-taking, and for many of us, speed it up. But perhaps the slowness of handwriting is a blessing, allowing us time for more reflection and contemplation. When written words keep up with thoughts, do thoughts stay at a shallow level? And if the words keep up, there is no time left over for contemplation. And what about the annoyance experienced by the people around us, for they too are trying to enjoy the event. The spoken voice, no matter how softly produced, will surely annoy, as will the clicking of the keyboard or the inevitable beeps of the electronic artifact. There is much virtue in keeping things simple, perhaps by sticking with old-fashioned technology. The technology must fit the pace and nature of the task and situation.

Why Bother?

With all the problems of recording events, why bother? I can see many reasons for recording an event for future use, whether through obstructionist, intensifying, or vicarious means. I am not opposed to records, for I too enjoy reminiscing about the past, sharing photographs of special occasions at family gatherings. Photographs allow a sharing of the experience, a sharpening of the memory. The trick is to have gotten the photographs naturally, without interfering with or even destroying the original event. But today, the recording of events has taken on its own life, seemingly performed solely for the sake of the recording itself. Many of the instances that I see just don't make any sense.

I always wonder about all those tape recorders facing me in the lecture room. If the students don't have time to come to class, why will they have time to listen to me later? After all, listening to a recording takes just as long as listening to the original. Yet lots of people who don't have time to come to the original event record it instead.

The ultimate folly is when people tape television shows, meetings or family gatherings, or have other people tape them for them, but then, having taped them, never watch them. Ever. Maybe they think there is no need to watch them: Knowing that they are there is enough. It's like owning an encyclopedia. It's very reassuring, sitting there impressively on the bookshelf. You don't ever actually have to read the encyclopedia: Just having it is enough.

Or, what about students who both attend my lectures and record them? Gad, it's bad enough to have to suffer through me once without having to do it twice. What will be in that tape that is so precious? Actually, the second listening can actually take longer than the original. True, in listening to a tape you can skip over the bad parts, but only if you know beforehand that it is a bad part. And if you try to listen to any section carefully, especially if you want to take detailed notes, it is apt to take considerably longer than the original experience.

In many areas of science, especially social science, it is necessary to create typed transcriptions of tape recordings of speech. Those who create the transcriptions—typing each word as it was uttered with high precision—require as much as ten hours to transcribe a single hour of audiotape. Try to transcribe a video recording and the time goes up immensely. Twenty, thirty, forty hours per hour of tape is not uncommon. Why? Because those who record have to experience and record every pause, um, hmm, and cough, every glance of the eye and false hesitation, any one of which may be important for the understanding of the event. And it usually is.

Human communication is a rich process. We do not communicate by words alone. The written word is a different thing altogether than the actual speaking, gesturing, moving presence. In spoken language the subtleties are conveyed both by speaker and listener, the one clueing the other as to the amount understood and the points of real interest. Written language makes up for this by relying on many artificial conventions, which is one reason why written language is so hard to produce.

If you want to chat with me for an hour, all we need is an hour's time. We arrange to meet. We chat. And that is that. If you want to read my words for an hour, then I must work much harder. In fact, because you can read between 100 to 300 words per minute, I must prepare 6,000 to 18,000 words of text: from thirteen to forty pages of printed text to fill your reading hour. That may take me ten hours of writing (and this would be considered very fast writing indeed). That much material could easily take a month to write, longer if the material was at all complex.

Notice the discrepancy. It takes about ten hours of work to transform an hour's worth of spoken speech into a typed, word-for-word transcription. And it takes about ten hours of work to transform the ideas in my head into a text that you can read in about an hour.

Written and spoken speech are so different that we are ill-served by artifacts that too readily attempt to convert one into the other. The difference becomes dramatically apparent if you ever read a transcript of a spoken interchange. What appeared to be fluent, graceful, profound speech in reality turns out to be clumsy, repetitive, and ill-formed in the reading.

I wonder at the thousands of dollars conference organizers sometimes spend to tape-record every "precious" word of a conference. Sometimes a crew of three people is required to work the microphones and tape recorders, to keep a written record of which tape goes with which session, and to ensure that a fresh tape is always ready when the old one is full. The more sophisticated crews use several tape recorders so that no gap occurs in the recording when it is necessary to change tapes.

But for what purpose are these recordings made? Why bother? In all the conferences that I have attended where recordings have been made, mostly they sit unlistened to when it is all over.

The effort of listening to hundreds of hours of tapes is overwhelming, especially when the transcripts will miss the subtleties of live interaction and provide instead deadly, stilted, halting words.

Life Through Videotape

My scientific research focuses upon how people interact with technology. Many of the interactions are recorded on videotape so that the interactions can be studied in detail. As a result, I often spend hours at the university watching videotapes, carefully backing up after critical events and reviewing them, the better able to understand each word or observe each movement. After a day of this I come home to watch the evening news. When I miss some critical phrase or event, I quickly reach for the control and try to reverse the action. But I can't: Real life just can't be manipulated as easily as tape. Or can it?

Consider the extreme case of a recording and playback technology run amok. A real meeting must be endured for its entire duration. A videotape of a meeting can be viewed selectively, with individual segments ignored, skimmed, or listened to repeatedly. Suppose this were possible with real life: Would that be a blessing or a curse?

Suppose I were to have a tape loop always at work on my home video recorder. It is a loop of tape that lasts, oh, three hours, with the current events always being recorded over the section of tape that happened three hours ago. I watch the TV, not live, but on tape, just delayed a fraction of a second after the events really happen.

Now suppose I come home tired after a day at work and want to watch the evening news. If at a critical point I have to leave the set, I simply put the world on hold. The recording continues, but the playback head marks the place on the tape where I stopped watching. When I return to the room, to resume watching, the playback continues from exactly where I left off. As far as I am concerned, I am watching the news as it happens, but in fact, I am watching the events delayed by the amount of time I was out of the room. (Figure 1.1.)

Suppose I am watching some sporting event and I want to review the last play: I flip my controller to reverse and back up to the point where the interesting event starts. Now I can watch it with more care, reviewing it as much as I like. Then I can resume the

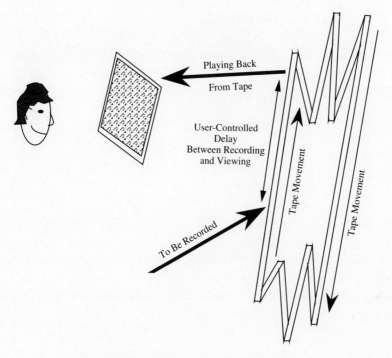

Figure 1.1 The 3-Hour Tape Loop

playback of the sporting event. Once again, as far as I am concerned, I am watching the event as it happens, but in fact, I am watching the event delayed by the amount of time I spent playing back the interesting parts. If something is boring, I can zip ahead as fast as I like (although I can never get ahead of the recording head, or in this case, of the actual events).

Now consider the Event Fanatic of the Future—call this fanatic "Eff." Eff views the real world through a TV lens, and listens to it through earphones. TV goggles are securely strapped to his head, electronics are strapped to his waist, lenses and microphones are mounted on his head. At first glance, this would appear to offer Eff no advantage: He simply sees everything through TV instead of live, like the rest of us. But Eff views the world through a tape delay. When something boring happens, Eff can fast-forward through it, at least until the tape is synchronized with reality once again. When something interesting happens, Eff can repeat it, right away, secure in the knowledge

Figure 1.2 Eff, the Event Fanatic of the Future, experiences the world through lenses and microphones on "goggles" that record scenes in the real world, pass them to the electronics belt on his waist, and then return them to miniature TV screens inside his goggles. Interesting scenes can be stopped and reviewed at leisure. Boring scenes can be turned off, or even skipped via a fast-forward control.

that if anything else interesting was happening at the same time, it will still be available for watching.

Pity the poor professor lecturing to Eff's class: Eff is paying careful attention, but it may be to something that was said five minutes ago. Or maybe to a current soap opera—who can tell? But then, in this case, maybe the professor is replaced by a computer-generated television image. Artificial images teaching artificial minds.

Alas, we seem enamored with our recording technologies, so much so that the act of making the recordings dominates. Will we also become enamored with playback technologies? Will the world be filled with cvcnt fanatics, lacking patience with real life, always wanting to skip around, now skimming ahead, now reviewing the past, continually skipping from one source of material to another?

Experience with technology teaches us that once a technology makes something possible, it gets applied, whether for good or bad. It makes sense to be able to show a sixth grade play to interested rela-tives—grandparents, perhaps—who could not attend. It makes no sense to destroy the experience through the act of recording it. It makes sense to have control over the viewing of records. It makes no sense to sacrifice human social relations in the process. Which will it be? I put my faith in people. Human social interaction is too impor-tant, too fundamental, to fall to obstructionist artifacts and event fanatics.

2

Design Follies

EVER since I wrote *The Psychology of Everyday Things*, I have been persecuted by design, especially the follies of design. Actually, even the title of that book is a design story. *The Psychology of Everyday Things*: a neat, nice-sounding title with the clever initials "POET." All my academic friends liked the title.

Not so for the readers. With that title people interested in psychology avoided the book because they cared about people, not things. People interested in things—designers, for example—never heard of the book because it never occurred to them to look under psychology. Mail and comments from readers convinced me that the title should be changed. In fact, the paperback publishers insisted on it. The paperback version is *The Design of Everyday Things*—in retrospect, a much better title, even if its initials are the nonsensical "DOET." A good rule of thumb in design is that if the designer really likes some special feature, it should probably be the first thing to be discarded. This turns out to be a useful rule in writing as well: The writer's favorite passage is probably the one that should be deleted. Well, the rule clearly applied to my own production, the title of that book.

In POET, I discussed design problems with light switches, water faucets, and doors at great length, so much so that I saw no need to repeat any here, even though examples continually arise to amaze me. I thought I had completely sworn off doors, until I saw the Drabble cartoon included in this chapter. The "hero" of the cartoon strip is named Norman (a fiendish coincidence), so let me rush to Norman's defense.

16

Figure 2.1 DRABBLE reprinted by permission of UFS, Inc.

Why does a simple thing like a door need an instruction manual in the form of the words "push" or "pull"? If the door were designed correctly, the instructions wouldn't be necessary. There would be nothing else you could do except the correct operation.

Look again at the door poor Norman is pushing against. The door hardware is a flat plate, raised slightly from the glass door. But flat plates are clear signs of pushing. I don't blame that Norman for pushing; this Norman would have done the same thing. In fact, I frequently stand quietly near doors, watching to see how people deal with them, and I find that people naturally push against flat plates. Isn't that why they are flat?

Door handles like the one tormenting Norman are clear examples of lazy, incompetent design. That's why they have to be accompanied by a sign that says "pull"—the flat plate is such an invitation to push that you have to warn people not to. People shouldn't have to read a manual to open a door, even if it is only one word long.

This chapter provides a sample of clumsy design practices found around the world. I have taken to traveling with camera, always on the lookout for yet another folly, a design that goes against its intended purpose, frustrating the user and often defeating the very purpose of the device.

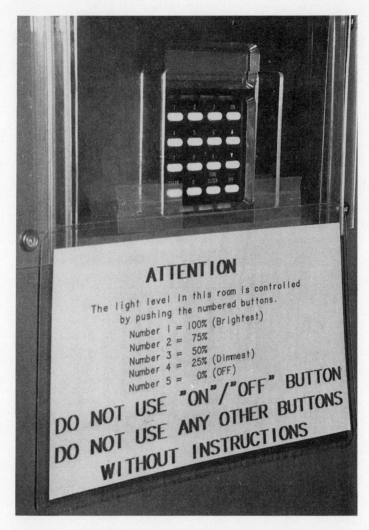

Figure 2.2 At a scientific meeting I attended, the speaker wanted to turn off the lights in order to show slides. Hah! This is what confronted us. Now imagine what happened when the slides were finished and we wanted the room lights on again. How could we read that sign in the dark so we could know which button to push? For that matter, how would we find the button? (We simply hit the "off" button and when the slides were over, poked around until the room lit up again.)

I start with examples of common, everyday items that ought to be usable without any instructions. Yet each of these devices required a large posted notice or a note, put on after the fact not by the designers, mind you, but by the frustrated owners of the devices who obviously tired of having to tell people how to use these everyday devices.

Who would think of needing an instruction manual to open a door, to control room lights, to use a water faucet, to work a public drinking fountain, or to open a chest of drawers in a hotel room? Why would anyone design an everyday item requiring instructions for its usage? Well, not only do designers design things this way, they can be found everywhere. I've included some pictures here as evidence.

I have a rule of thumb for spotting bad design: Look for posted instructions. Whenever you see signs explaining how to use something, it is a badly designed piece of equipment. I take great delight in traveling the world, reading all those taped, hand-written signs that instruct folks what to do and, of equal importance, what not to do.

A challenge to the designers of the world: Make signs unnecessary.

Tales of Affordances

Affordances is a strange word, a technical term that refers to the properties of objects—what sorts of operations and manipulations can be done to a particular object. A door affords opening and closing, a doorway affords passage. A chair affords support, which means we can stand, sit, or place books and papers on it. Tables also afford support, but because of their height they may or may not afford easy sitting. A ball—or any similarly sized, movable object—affords throwing. If the ball is made of a hard, dense material, it also affords hammering, smashing, or mashing. It also affords rolling. Flat horizontal surfaces afford support, flat vertical surfaces afford pushing.

Affordances play important roles in many of the devices that give us trouble. In the design of objects, the critical thing is "perceived affordances," what the user perceives can be done. It is through perceived affordances that we know how to work devices we have never encountered before.

When we first encounter something, how do we know what it is used for, and then, how to work it? The answer is complex, but

Figure 2.3 It's fun to watch people try to get water out of automatic faucets like this. Many people have discovered that if they lift up hard on the bottom of the spigot, the water will come out. Clever. Lifting up hard gets the hands under the tap, which engages the infrared sensor. The lifting itself has no effect. Automatic faucets are useful, sanitary, and water-saving, but the sensors should be more visible so that users can recognize them.

our perceptions of the object's affordances play a major role: If something doesn't look movable, we probably won't try to pick it up, or throw it, or move it out of the way. We push things that appear to be pushable, pull those that appear pullable. We work best those things that have only a single, unambiguous, clearly perceived affordance: After all, there is only one possible thing to do, the right thing. Objects that have no relevant affordances are hard to work (a principle used effectively in designing those puzzle boxes that appear to have no way of being opened). Objects that have too many affordances are also difficult, for the user doesn't know which of the several possibilities to select.

Perhaps the worst case of all, however, is when there is a single, clearly perceived affordance, but it is for the wrong operation. This is the plight of poor Norman in the cartoon sequence. The door of his encounters has a flat plate as its hardware and flat plates clearly and unambiguously afford pushing, which is what Norman was doing. In

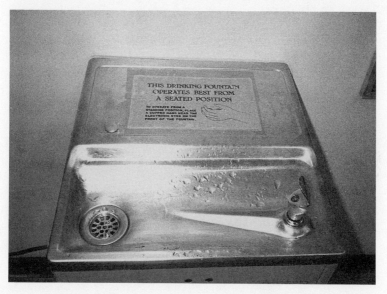

Figure 2.4 One day the drinking fountain shown in (a), above, appeared in the undergraduate library at my university. People couldn't figure out how to get the water out of it. It had hidden sensors (the two dots on the front—not very obvious and not even visible from a normal standing position above the fountain). Eventually, the library staff had to put up a sign (b), below. No obvious affordances. The fountain is designed for people in wheelchairs, so that when they wheel up, the water will start automatically. As in Figure 2.3, these sensors have to be visible if we are to know that they are there.

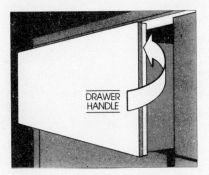

DRAWER
HANDLE

Figure 2.5 Invisible affordance strikes again. You can see why instructions are needed for these dresser drawers: There is no obvious way to get at them. Even with an instruction manual, these drawers are not easy to open. They do look pretty, however.

fact, it is the proper hardware for a door that should be pushed. Alas, we often find plates on doors that should be pulled, which is why Norman (and so many of us) push when we should pull, which is why so many doors have to have instructions on them. Bad design has to resort to signs or instruction manuals to overcome the design flaws.

Why do people design such things? To a large extent because they lack empathy with the users of their devices, as I discuss in Chapter 17 ("Writing as Design, Design as Writing"). A good designer has to take the user's point of view, consider what information the typical user must have to work it properly, easily, efficiently. The best way to do this is to observe and interact with the typical people for whom the product is designed. This means testing out the designs on them—and being willing to change them when users find themselves confused. Alas, most designs concentrate on other things: on cost, or aesthetics, or technical details, not on how a first-time user can figure out how to operate a product. I suspect most designers have never actually watched people use their products. As a result, users get confused, and the signs soon appear—instructional signs, warning signs, desperate signs.

A second reason that we have such problems is aesthetics, a mis-placed aesthetics, of course. Ah, what we must put up with in the name of aesthetics. Look again at the chest of drawers in Figure 2.5. If the drawer simply had visible hardware, the user would know what to do! (The same comment applies, of course, to all those sleek kitchen cabinets that offer no hint of where to open them, or even which side of the door is hiding the hinge that opens the door.) Ah yes, the de-signer was concerned that the hardware would have detracted from the clean, sleek lines of the drawers. Leave out the hardware, however, and the first-time user cannot see how to open them. And hotels, of course, are flooded daily by visitors, most of whom will be first-time users of these drawers. Finally, another consequence of poor design rules: The sleek aesthetics are ruined by the posted notes explaining to the person how to use the stuff.

If the lack of appropriate perceived affordances can sometimes cause trouble, affordances that send the wrong message can cause equal trouble. We have already seen one example of this in the flat plates that signal doors to be pushed when in fact they are to be pulled. Here is another.

Figure 2.6 How can you get ice cream out of this freezer—while holding a full tray with one hand? You can't. You need to put your tray down. Where can you put it? The only spot is the right-hand top of the freezer, while you open the left-hand top and reach in. But the top wasn't made to support any load, and eventually broke. The sign was too late. Bad affordances; bad cafeteria design. (Why didn't they provide any other place to put the tray?)

At one of the cafeterias in my university, I noticed a food freezer containing ice cream positioned along a wall near the cash registers. Customers pick out whatever ice cream they wish, place it on their trays, and pay for it when they go through the checkout lines at the cashier. The freezer is a long, horizontal box with solid sides and a transparent glass top. The top can slide open so that people can reach in to get the item they want.

But on top of this freezer is a sign: "PLEASE DO NOT PUT YOUR TRAY HERE!" Aha! A nice flat top—a clear signal of support, the affordances obvious. Moreover, if you have a full tray, how can you lean over the freezer and reach in to get some ice cream without spilling the contents of your tray? You have to put the tray down somewhere and the only logical spot is the top of the freezer. Clearly, the affordances are all wrong.

After I first encountered the sign during a lunchtime excursion to the cafeteria, I made a special trip back, camera in hand, to photograph it. Too late. The top had already been cracked and a worker was cleaning up around it. I asked what had happened. "Oh," said the worker, "It broke. Too many people were putting their trays on top." As a result, my photograph shows not only the sign but the crack. Signs don't work. People's needs and perceived affordances will dominate every time. The proper solution would have been to do two things: (1) satisfy the need for a tray holder while getting out the ice cream, and (2) change the perceived affordances of the top. (Put something in the way, make the top slope, do something to change the perceived affordances so that it no longer looks like a place that can support trays.) Actually, I prefer a more humane alternative: Admit that the top is an excellent place to hold the trays and make it strong enough to do so safely.

Whatever the solution adopted, if it doesn't properly address the needs of the cafeteria users, it will fail to be effective despite any warning signs and admonitions.

Tales of Mappings

Affordance is a critical design concept, mapping another. Mapping too is a technical term and, like affordance, an important one. Mapping, as used in design, refers to the relationship between different parts of a system, or between the controls and the results. Suppose you have

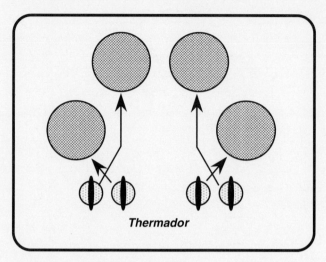

Figure 2.7 A stove that could have gotten it right but failed because Thermador mismapped the relationship between controls and burners.

four burners on your stovetop and four controls to operate them. The correspondence between the controls and the burners is a "mapping." There are good mappings and bad ones. Good ones are what I call "natural mappings," where there is an obvious, spatial relationship between the layout of the controls and the things being controlled. Needless to say, most designs seem to get it wrong. And nothing demonstrates it better than the everyday kitchen stove.

The major problems with stoves and their controls is that most stoves have four burners arranged roughly in a rectangle, with the controls arranged in a straight line. As a result, it is impossible to find a natural mapping between the control arrangement and the burner arrangement. Worse, there is no standardization, so that the system used by one stove is apt to be different from the system used by another.

I thought that I had thoroughly documented the foibles of the kitchen stove in POET and that there was no more to say on the topic. But the designers of stoves have outdone themselves. There is a new generation of designers who have done it even worse.

A reader of POET sent me a photograph of his American-made Thermador stove that staggered the four burners so that they formed a

26

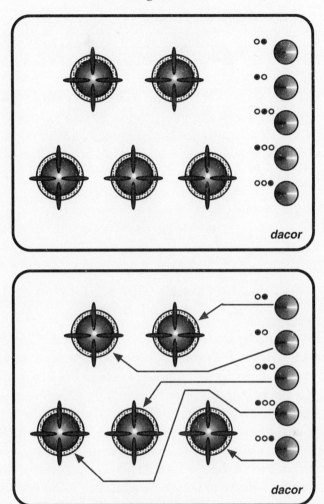

Figure 2.8 Five burners, five controls. Look at (a), above: Try to figure out the rationale for the relationship between the burners and the controls. (The manufacturer, Dacor, tried to help with the little black dots in the circles just left of the controls.) In (b), below, I have added arrows to make the relationship easier to see, but no easier to understand. Why isn't the middle burner in the bottom row controlled by the middle of the three bottom controls?

nice left-to-right ordering. The four controls were arranged in a straight line, from left to right. Hurrah! With this design, the four controls from left to right could correspond nicely to the four burners, from left to right. But did Thermador take advantage of their nice layout? Of course not. The burners went one way, the controls another. The photograph didn't have enough contrast to be reproduced here, so I have drawn the ordering for you. The arrows, of course, are added by me and are not on the stove.

Or pity my relative who installed a lovely new stove in his newly redecorated kitchen. Five burners, three in front and two behind, as shown in Figure 2.8. Five controls (in a straight line, of course), separated into two groups, one with three controls for the three burners in front, one with two controls for the two burners in back. But what a mapping! Of the three front controls, for example, wouldn't you think that the middle control ought to control the middle burner? Hah!

The Sydney Monorail:
Bad Affordances, Bad Mappings

In Sydney, Australia, an elevated monorail train links part of the downtown area with the harbor and the convention center. When I visited Sydney in 1988, the monorail had been in operation only a few months. The equipment was still new and shiny. The system was very popular; people were proud of it. To enter the monorail required a token—a metal coin-like object specifically designed for the monorail. The token could be purchased by inserting coins or paper money into a token-dispensing machine. The token was then inserted into the token-accepting machine of the entrance turnstile, which allowed the rider to enter and provided a paper receipt. The token-dispensing machine and the token-accepting machine (the entrance stile) were both masters of modern technology. For example, the token dispenser accepted a wide variety of coins and bills, allowed the user to specify how many tokens were to be purchased, and then delivered both the tokens and the proper change. Each monorail station was intended to be highly automated, requiring a minimum of personnel.

Alas, when I visited, each machine had to have a full-time attendant standing beside it to help out the bewildered users. What was the trouble? Why couldn't these apparently well-designed machines be

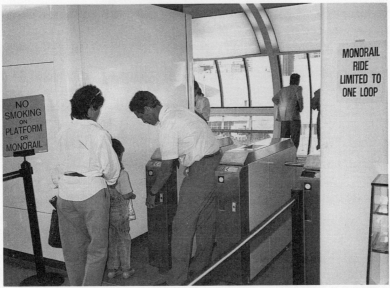

Figure 2.9 An automatic turnstile for the monorail in Sydney, Australia. Passengers tend to put their tokens into the slot that dispenses tickets because it is more visible than the proper slot. The added sign in (a), above,—the "X" over the most visible slot—is a vain attempt to prevent people from putting their tokens there. Note the need for a full-time employee (b), below, to guide passengers. The best solution would be to put a barrier in front of the ticket dispenser so that it is not visible to passengers until after they pass through the turnstile.

used without assistance? I asked this question of one of the attendants, patiently standing at the turnstile, showing people where not to insert the tokens. "It's because people can't read," she explained. "We've tried all sorts of signs and people still insist on putting their tokens in this slot, where the paper receipt is supposed to come out." "Hmm," I wondered, " this slot looks pretty big and obvious. It's the first thing you see when you look over here. Maybe that is the problem." "Well, yes," she admitted and thought for a moment. "Maybe if we put up a shield so people couldn't see the slot until they went through, maybe then they wouldn't make that mistake."

Yes, indeed. The design was all wrong. Look again at Figure 2.9: The first thing visible is the dispensing slot, and a slot affords insertion. No wonder people tried to put their tokens there. If that slot were not visible, the problem wouldn't have occurred. Of course, if the slot weren't visible, then after putting the token in properly, people might not notice where to get their receipt. But suppose the receipt-dispensing slot were moved backward to where it couldn't be reached before passing through the turnstile? That might do the trick.

It seems clear that the affordances were wrong on the device: There were clear affordances that conveyed the wrong message. As for signs? People rushing to catch their train, struggling with packages, companions, and children, are not very likely to look for and try to understand signs. Especially signs that should not have been needed.

When we ourselves passed through the turnstile, my son carefully inserted the token into the token-accepting slot, but for the wrong turnstile. He was puzzled when he couldn't get through the one he was pushing against, but the vigilant operator had noticed the problem and led my son around and through the appropriate turnstile. "Happens all the time, it does," the operator announced to me. Ah yes, the mis-affordances of the token slot were coupled with an unnatural mapping of slots to machines.

What about the token-dispensing machine—why was a full-time attendant needed? Bad mapping, that's why. The first thing that strikes the eye upon walking up to the token dispenser is the slot for the money. So that is where the action starts. Not only that, but that is an obvious way to begin. Look carefully at the signs on the token dispenser in Figure 2.10: The instructions are labeled 1, 2, and 3, run-

Figure 2.10 The automatic token dispenser serving the Sydney monorail. Despite "automation," a full-time attendant is required to assist users. The steps required to get the token from the machine are labeled 1, 2, and 3 on instructions in a left-to-right order. The actual operations start at the upper right (the round button), then the upper left (the coin slot), then down below (the output slots). Customers (sensibly) also try to do the operations in a left-to-right order, putting the money in first (which also seems like a logical way to start). Ah, but the machine wants this to be step 2, not step 1.

ning neatly from left to right. But the operations to be done do not follow this same order. Operation 1 must be performed on the right, operation 2 on the left. Why aren't the operations also in a left-to-right order, perhaps with the appropriate place just under the elegantly labeled instructional sign? But why is the machine so fussy about order? Why not design the system so that it can accept either the money first or the specified tickets and destination first? Order shouldn't matter as long as the amounts and requests correspond; if they don't, the user can be told. After all, if a person instead of the machine were selling tokens, the person would allow for flexibility. Why can't the machine be just as flexible?

It is rare that any single design can be so wrong in so many places. The Sydney Monorail "Automated System" is a remarkable experience.

Sins of the Future

I have seen some of the technology of the future and it looks suspiciously like that of the past, except with fancier buttons and displays, and more elaborate hardware. Look out folks, here comes the Home System, computer-controlled appliances, all connected together. Just imagine, you will be able to turn on your bathtub while driving home after a long day's work. The same designers that brought you the videocassette recorder that you couldn't program, the digital watch that you couldn't set, and the stovetop controls that you couldn't learn will now bring you the fully automated house. Would you believe a bathtub that requires an instruction manual?

A few years ago, a conference of the International Congress of Psychology was held in Sydney, Australia. One of the nice things about international conferences is that they are held in different countries, so attending them is both a way of keeping up with scientific developments and also of seeing the world. I have already discussed the monorail system I encountered in Sydney.

From Sydney, my family and I went to Brisbane to the International World Fair (EXPO-88). EXPO-88 had the typical set of exhibits, with many nations having special buildings to show themselves off, and a few individual organizations launching separate exhibits on their own. I spent only one day there, not enough to cover all the exhibits. Many used high technology, either as information aids to help the fairgoers find their way or to inform them about the wonders of the nation or organization hosting the display. Some of these worked well, others were frightfully complex, serving mainly to frustrate. A few exhibits tried to predict the future, especially the way in

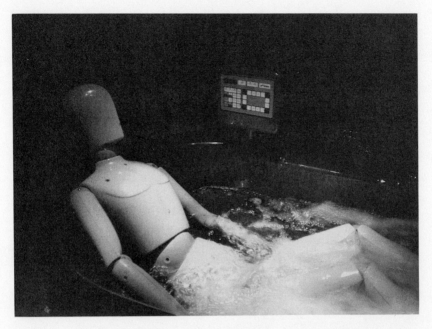

Figure 2.11 Italian high tech—the automated bath rub. Now you will need an instruction manual to take a bath! Seen at Expo-88 in Brisbane, Australia.

which technology would affect everyday living. For me, the highlight of these exhibits had to be the ultimate folly in home automation, the automated bathtub.

The automated bathtub comes complete with a complex, illuminated control panel. I never did quite figure out what all the controls for the bathtub were supposed to do. My bathtub has four controls: one for hot water, one for cold, one to engage and disengage the plug allowing the tub to fill or empty, and one that determines whether the water comes out of the spigot for the tub or the shower. Four controls seem quite enough.

The high-technology bathtub has approximately thirty push-button controls plus a numerical display. I never figured out what all the buttons did, or why the lovely nude mannequin in the tub never tried to operate the controls. But as you can see in Figure 2.11, the bathtub was a marvel to behold. Italian high-tech. As the conference brochure for the Italian pavilion explained, "Leisure should not be considered a second-rate experience in a life totally dedicated to work and domestic obligations. . . . The pavilion illustrates a house in which the latest technology makes living easier." Sure it does. Now I need an instruction manual to operate my bathtub. Ah, for the good old days, when all one had to do was to put in the plug and turn the water on.

However, the prize in design folly must go to the U.S. Post Office for its new stamp machine. I won't go into detail about the design sins of this beast—its inappropriate mapping, its lack of informative feedback, and its other design flaws. Suffice it to say that I have seen people get emotionally upset as a result of their interactions with this machine.

Remember my rule of thumb that when a machine requires an added-on instructional sign, it indicates poor design? Well, the machine at my post office (in Del Mar, California) has not only handwritten signs on it but a fancy, computer-controlled sign with scrolling red letters that says:

WELCOME TO THE DEL MAR POST OFFICE VENDING MACHINE *** I REFUND A MAXIMUM OF \$3.25 CHANGE WITH YOUR PURCHASE *** THINK BEFORE DEPOSITING A BILL LARGER THAN \$5 ***

Now put yourself in the place of the patron who has just inserted \$30 into the machine in order to purchase a roll of 100 29-cent stamps, expecting to get the stamps and \$1.00 in change. First, the machine graciously accepts your money, and then it informs you that it no longer has any 29-cent rolls. What would you like to buy instead? And no, it won't return your money (no more than \$3.25, remember?). Yup, it happens.

This design is so bad that they had to design special signs to warn people. There is only one thing more suspicious than a handwritten warning sign on a device and that is a professionally printed or prepared warning sign. The latter means the problem is so severe and frequent that it was worth someone's time and effort to prepare the warnings.

How can you trust a vending machine that warns you to "think" before you can use it?

Design Gems

It isn't fair to discuss only the negative side of design; so let me tell you of some examples of pleasant, elegant design, design that informs, that is usable, and that improves upon life rather than detracting from it.

I have come across a large number of very innovative designs. A clever use of computer technology at the post office in Brisbane, Australia, made it easier for people in line and clerks alike.

Remember how horrible it is when you have to decide which of several lines to stand in? No matter which one you choose, yours is the slowest, right? It feels quite unfair to wait in line for a long time only to see others who arrived after you get served first simply because of luck in the choice of lines. A standard major design improvement in line-forming has been the development of single lines for multiple clerks. The single line lets customers be served in the same order they arrived in line by whichever clerk is next free. This makes the waiting experience much more equitable. But when there are many clerks, sometimes it isn't easy to tell when one is free. The Brisbane post office solved the problem by mounting a sensor under the carpet in front of each clerk so that it can automatically sense when the previous customer has walked away and the position is free. A computer then flashes the number of the free clerk on a screen in front of the waiting customers, along with an arrow pointing in the proper direction to walk. I asked a clerk what would happen if she needed to take a break, or if all the windows were not being serviced. The clerk showed me a switch that would simply disconnect the window position from the list. The entire scheme was very clever: It used a relatively simple technology to improve service for customer and clerk alike.

A second example is the deliberate use of misleading affordance at the Asilomar Conference Center in northern California. The center wished to close off some roads to tourists and guests but still permit passage by service vehicles. So deliberately deceptive anti-affordances were employed—posts that appeared to block the road but that in fact bent easily, permitting passage to those vehicles that knew the secret. The deception worked. Service trucks simply drove right over the

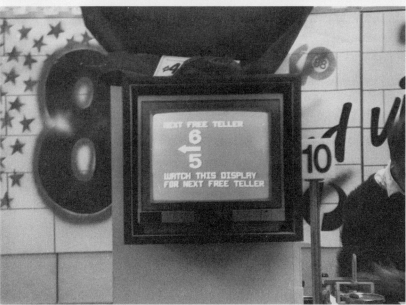

Figure 2.12 An automatic line controller at the post office in Brisbane, Australia. Customers form a single line for multiple clerks. A sensor under the carpet tells the system when a customer is present at a window. The TV monitor tells customers when and where to find an empty window. Very simple, clever, effective.

posts while tourists and lodging guests, thinking the posts were as solid as they appeared, didn't attempt to enter. Here is another clever use of simple technology. The aquarium in Sydney, Australia, placed television cameras in front of the tanks displaying small marine animals and plants. The aquarium visitors controlled the cameras, deciding both where they were aimed and the magnification (the "zoom" setting). This let the visitors see magnified images of the marine life on the television displays. A clever use of technology to make visible what would otherwise be invisible.

It doesn't take very much technology to be of great assistance. I was very impressed with a simple mirror in the Cyprus Gardens amusement park in Orlando, Florida. That's all it was: a large mirror. You know all those vacation pictures that have the whole family in them except for the person taking the picture? Well, the mirror solves that problem. It lets tourists view themselves—and more importantly, take pictures of themselves, with the camera person in the picture too.

Another innovative, clever design is the clearly marked, very visible "Meeting Point" by the baggage claim area in Amsterdam's Schiphol Airport, a boon to travelers trying to find relatives or friends. Or the Canon laser printer, constructed so that deep inside, where there are numerous connectors for its electrical cables, each pair is unique, different from all others. This is a very important, very clever safety feature. It means you can't put it together incorrectly. If only other industries would practice this philosophy, we would have fewer errors in the assembly and maintenance of equipment, errors that have led to serious injury and death.

Alas, poor design seems more prevalent than good design. In general, design failures are caused by lack of thought, not by malicious intent. Automobile radios that cannot be operated safely while driving. Flap and landing gear levers on private aircraft that look the same and are located next to each other. Ford has a foot brake release and hood release that look the same, and are positioned next to each other. Yup. You meant to raise the flaps and instead raised the landing gear? A very expensive accident. Meant to release the automobile's brakes and the hood pops open? Lack of thought can be dangerous.

There are good design principles that avoid these problems. In all commercial airplanes, not only are the flaps and landing gear controls separated from each other but a principle called shape coding is used. The flap control is shaped like a flap—a flat horizontal surface, rounded in front, tapering toward the back. The landing gear control is shaped like a wheel. If the wrong control is grabbed, the error can be detected immediately by the feel. A very simple, elegant solution, well known for close to fifty years. Yet another design gem.

By now you should have gotten the point. Moreover, you should be ready to look around you, to spot those poor, inhumane designs, and to complain to the companies that force them upon us. But do not forget to applaud the design gems, the ones designed to serve, to enhance, and to improve our lives.

Happy exploring.

3

The Home Magazine Kitchen

ARTICLES and advertisements in home magazines always show bright happy families in bright happy kitchens. They excite the imagination, but not always in the ways the magazine intended.

How come there are never any:

> Dirty dishes?
> Dishes in the drainer?
> Drainers?
> Spills?
> Clutter?
> Long rows of appliances?

And how come there always seem to be enough electric outlets?

Every year the busy manufacturers of the world toil and labor to produce yet another kitchen appliance. Would you believe that even now, deep in their secret caverns and laboratories, scientists are plotting to take over the kitchen with their diabolical means?

The intelligent home system is coming. Ah yes. We will have communication lines connecting all our appliances. The refrigerator will talk to the stove. The stove will talk to the washing machine. The washing machine will talk to the heating system. All the appliances will talk with one another. And the worst part of all is that they may try to talk to us.

"Hi, Don. This is Fred, your friendly dishwasher. I am ready to do a load of dishes. Would you like me to wait until 3 AM, when you are asleep and when utility rates are low?"

"Don? Don? Don, I am talking to you."

"Hi, Don. This is Fred, your friendly dishwasher. I am ready to do a load of dishes. Would you like me to wait until 3 AM, when you are asleep and when utility rates are low?"

"Don? I know you are there, Don. I can sense your infrared emissions."

"Hi, Don. This is Fred, your friendly dishwasher. I am ready . . ."

Hour after hour after hour. And then Susan, my friendly refrigerator, will ask if she can defrost herself at 4 AM. And let's not forget Tom, my friendly oven. Oops, Tom is also my friendly videocassette recorder. Uh-uh, those manufacturers didn't synchronize their naming practices. What happens if I ask Tom to turn on from 2 AM to 4 AM for channel 10?

"This is Tom, your friendly oven. What temperature should I use?"

Or what if Susan asks if she should defrost the chicken, but before I can say "No," Tom says, "Yes, I am ready to record." Susan hears the "yes," so she defrosts the chicken. And maybe Fred starts the dishes.

Ah yes, the high-technology house. The same folks who brought you the digital watches and VCRs that you can't program, who brought you washing machines and dryers with unintelligible dials, and refrigerators whose temperature controls you can't set will now bring you the high-tech house. Welcome to the world of the future.

The Four-Questions Test

Whenever I see photographs of kitchens in those home magazines, especially ones that look appealing, I peer carefully for any sign that the kitchens are actually used. I look for answers to the questions I have about my own kitchen. I never find them.

Each appliance accomplishes some wonderful task that you never realized you needed but that now you will not be able to live without. How can you ever resist? Well, my family has figured out how. We invented a test—the four-questions test—that each prospective new purchase must pass. We don't care how valuable, useful, efficient, or essential the new appliance is; it has to pass this test or we will not buy it.

Question one, the most basic of all, defeats most new appliances right from the start: *"Where would we store it?"* Since we can't even figure out where to store the appliances we already own, what would we do with new ones? Put them on the floor? In fact, in my family this issue alone has become so serious that it has almost put an end to buying. I hope.

Now for question two: *"Where would we use it?"*—even if we could find a place to store it. Wherever we try, there is something else already there. Look again at those magical kitchens in the glossy magazines. Where are the mixers, toasters, blenders, grinders, coffee makers, knife sharpeners, pencil sharpeners, juice makers, . . . , what-have-you?

Question three follows logically: *"Where would we plug it in?"* The real trick is to find an available outlet that has some relationship to the place where we would like to use the appliance. These new kitchen gadgets always require plugging in. I think that if we bought another breadboard, it would require plugging in. Either that or it would use batteries—and, of course, it would require some assembly.

Question four brings up a critical issue: *"How much work would it be to clean?"* We have managed to escape several appliance fads simply by thinking about the cleanup problem. Sure, a kitchen atomic vaporizer will save us thirty seconds every time we need to do some atomizing, but at a cost of five minutes of disassembling, washing, drying, and reassembling. If we could figure out where to put it, and use it, and plug it in.

Designing Kitchens for Real Life

I suspect that kitchens are designed to look good, not to be used. If kitchen designers had their way, they wouldn't let ordinary people into kitchens, only photographers. The problem with kitchens is a general problem with design. How many people really worry about the *usability* of their kitchen—or their home or office or computer or the ever more fancy gadgets in their automobile?

The proper way to design something is to take into account what people really do and then construct things appropriately. Sound logical and sensible? Yup, but it is amazing how infrequently such common sense is followed.

Industrial Psychology and the Kitchen

Kitchens were actually one of the very first workplaces to receive careful, scientific study. In the early 1900s Lillian Gilbreth, one of the world's first industrial psychologists, studied work habits in home and industry. Among her many activities, she taught industrial psychology as a professor at Purdue University and did scientific studies of work patterns in kitchens. Gilbreth may be familiar to some of you as the mother of the family in the books and films *Cheaper by the Dozen* and *Belles on their Toes*. She and her husband, Frank Gilbreth, did indeed have a dozen children, and they practiced and polished their studies of industrial efficiency on their own family. It was two of the children who later wrote about their life.

Lillian Gilbreth studied the work patterns in the kitchen. She and her associates watched and measured, timing operations with a stopwatch—the Gilbreths were one of the pioneers in the development of "time-and-motion" studies, especially through the use of motion pictures of the activities. From this work, Lillian Gilbreth developed the concept of the "work triangle"—the proper arrangement of sink, refrigerator, and stove to make the normal activities of usage more efficient and less tiring. Gilbreth made sure there was a place for every object and every activity, a place that was carefully selected to minimize effort and increase efficiency and ease of use. To my knowledge, no equivalent studies have been conducted in recent times.

It is about time we updated Gilbreth's principles. Work patterns have changed dramatically in the kitchen since those times. Just as today it is rare to have a family of fourteen, other changes have come about in the home. For one, the kitchen used to be populated mainly by the cook, whether a member of the family or hired. In Gilbreth's own family the cook would not let other family members into the kitchen. Today almost everyone invades the kitchen, and kitchens need to be designed to accommodate them. Sometimes this population pressure is deliberate, as when two or more people collaborate in preparing a meal. Sometimes it is accidental, as when family members enter the kitchen for this or that while the main food preparer is at work. Are kitchens designed for these practices? Not very often. Moreover, we have many new appliances and completely different kinds of foods today: supermarket packaged foods, foods meant for the microwave oven, and foods that are frozen, canned, dried, and otherwise specially prepared. There are new kinds of cooking devices, new kinds of blenders, grinders, and mechanized appliances, and new cooking practices.

Even with all these changes, Gilbreth's fundamental observations and principles still apply, and still are woefully ignored. Where do we find room for all our gadgets? Where is the proper workspace to support our activities? What design principles have been applied to the work patterns within the kitchen? These are the aspects of kitchen design Gilbreth emphasized—the aspects that are lacking today.

A good kitchen makes it possible for others to enter to get a snack, a cup of coffee, or a glass of water or milk without interfering with the cooking process. Or for two people to work at once without continually bumping into each other. Or for someone to answer the phone, or just to hang around to talk.

What about cleanup, perhaps the most neglected part of the kitchen: Are appliances designed so as to be easy to clean? Are countertops easy to clean, with enough room beneath them so that you can scoop crumbs into your hand? Is there sufficient room near the sinks and dishwasher to place dirty items in preparation for washing them? Can you open the dishwasher to fill it or empty it without blocking other activities in the kitchen? And what about newly washed items? Where do they go while drying?

Once more I look through kitchen design books in vain for some signs of a rack or holder for drying dishes. Nope. Actually, I have found some clever ideas from Scandinavia. That figures: Scandinavian designers have always led the way in combining functionality with aesthetics. Too bad others have not followed. Unfortunately, those Scandinavian kitchens seem better designed for small families and small apartments. They wouldn't work in my household.

Design Neglect

The kitchen is really the high-technology center of the home. It is easy to make fun of the kitchen, but it stands as a microcosm of a lack of consideration for others in the entire design community. Benign neglect was once a popular political concept: Deliberately leave something alone because it will probably be better off without any attention, without the type of aid that politicians typically provide.

Well, perhaps what we have here in the kitchen is Design Neglect. Notice that benign neglect failed as a political strategy and so too does design neglect fail as a design strategy. The kitchen—or any other workplace for that matter—will not be an efficient, convenient place to work unless someone spends time, effort, and thought at the task. Worse, if effort is put only into aesthetics or cost, the result is almost guaranteed to hinder usability.

Beautiful kitchen cabinets are nice to look at, but many do some impractical things—like hide the handles for the sake of beauty, which means that you can't figure out how to open them. Appliances have moved toward computer controls and touch-sensitive keyboards, so that now it is impossible to figure out how to heat a cup of coffee or make a piece of toast. Can you use your microwave oven in all its glory? Do you know how to program for defrost followed by a wait cycle followed by a cooking cycle? Can you even set the temperature of your refrigerator?

There are enough problems in the kitchen to fill a book. That's fine for me. I write books. But what about the user of the kitchen? In my earlier book *The Design of Everyday Things*, I talked about how difficult it was to adjust the temperature of my home refrigerator. The labels on the temperature controls and their actual functions were completely unrelated. I was not alone: Many people wrote to me saying that they had the same refrigerator and they too could not adjust the temperature. One person was even an engineer who worked for the same company that made the refrigerator.

And what about the need for electric outlets? You would think that any modern kitchen designer would understand the need for sufficient outlets, or for room to store all those appliances, a place to put them during use. No, I think that designers try to make everything look good. And outlets are simply not thought of as a part of the kitchen. Outlets are for the electricians to worry about. But electricians do what they are told to do, and if nobody tells them how many outlets to put in, they will follow the electrical code, which probably specifies two outlets per wall, regardless of whether it is a kitchen or a bedroom. Design neglect.

The Proper Way to Design

The proper way to design anything is to start off understanding the tasks that are to be done and the needs of the users. In a kitchen, don't start with the appliances and counters, start with the people and their needs. This is how all things should be designed, not just kitchens.

What do we do in the kitchen? It isn't hard to discover: just observe some families. Patterns probably differ depending upon the kind of family, but I suspect that there are not that many different kinds of usage patterns—a dozen perhaps? That wouldn't be too hard to study and catalog.

Here is what is required for my family, and maybe for many users of a kitchen. First of all, we use our kitchen for multiple purposes, sometimes several at the same time. One important activity is that of loading up the kitchen, bringing in newly purchased groceries. For this activity, we need places to unload the supplies and places for the storing. Cooking is the most obvious function of the kitchen, and this too has many different phases and activities, often spread out over considerable time. For these, we need places and equipment for food preparation and the actual cooking. We also need places for casual food preparation—for snacks—and places for food in the process of cooking, perhaps simmering, perhaps defrosting, or perhaps prepared ahead of time and now awaiting the final steps. An important but often neglected activity is cleaning up and taking out the garbage. Can we transport packages easily through the necessary doors? Finally, our kitchen has become the modern "communication and control" center of the home, the place where everyone congregates for information, talk, and activities.

All these different activities have to be accounted for, especially when performed by several people simultaneously. One of the goals of a successful design should be to make all the activities go smoothly, so that each of us is properly accommodated, so that we do not get into one another's way.

The Kitchen as Communication and Control Center

Where would the modern family be without the refrigerator? Its glistening sides provide an ideal location for the graffiti of living. What a serendipitous design accident. Refrigerator doors are made of steel, and steel is magnetic. Refrigerator doors are large, flat, unadorned. Those surfaces of white-painted steel cry out for a use, and a use has been found.

Sometimes I wonder which is the more important function of the refrigerator: storing the food on the inside or keeping the messages on the outside. The family refrigerator has become the major center of the home, in part through the magic of magnets. In fact, in the United States, the making of "refrigerator magnets" has become a minor industry. And where would we be without the most impressive invention of the era: Post-it Notes?

It all had to be accidental: I can't imagine anyone in the refrigerator business or the kitchen design business being clever enough to design the kitchen message board on purpose. Even in countries where the refrigerator door is not used for messages and announcements, families typically do have a need for some central message board. Everyone must know of this need for centers. Everyone must know except the refrigerator manufacturers and kitchen designers.

Designer kitchens, of course, do not allow for such an activity. Look again at those home magazines: Never would any appliance be sullied by a magnet or taped note. Moreover, if you follow the designers, you will pay twice or even three times as much money and buy a designer refrigerator with wood panels; you will pay twice as much in order to have a door that has no use at all, except, sigh, as a door. Pretty, elegant, but so dull.

The refrigerator door, quite by accident, turns out to be a nice, functional place for messages. In part this is because in many homes its location, not so accidentally, turns out to be perfect. Home design has evolved over the years. Today, especially in the United States, the kitchen is often the center of family activity and the refrigerator the center of the kitchen: well lit, central to the other activities.

46

I can imagine a typical kitchen inhabitant during a between-meals forage: glass of milk in one hand, cookies in the other, telephone between shoulder and chin, checking for scheduled events and messages on the refrigerator. Do designers have such a picture in mind when they design? This is real kitchen usage but it is never planned, never thought about: It just happens. But because it is so essential to kitchen life, it ought to be a major part of kitchen design. It should be thought about, incorporated into the kitchen plans. I fear that such practical uses will be destroyed by the prettiness of the home magazine kitchen. Aesthetics before function. Ah yes. The story of modern design.

4

Refrigerator Doors and Message Centers

Sometimes I wonder which is the more important function of the refrigerator: storing the food on the inside or keeping the messages on the outside. The family refrigerator has become the major center of the home, in part through the magic of magnets. In fact, in the United States, the making of "refrigerator magnets" has become a minor industry. And where would we be without the most impressive invention of the era: Post-it Notes?

It all had to be accidental: I can't imagine anyone in the refrigerator business or the kitchen design business being clever enough to design the kitchen message board on purpose. Even in countries where the refrigerator door is not used for messages and announcements, families typically do have a need for some central message board. Everyone must know of this need for centers. Everyone must know except the refrigerator manufacturers and kitchen designers.

The paragraphs quoted above are taken from the preceding chapter, "The Home Magazine Kitchen." The original essay had an extra paragraph, one that got removed during the writing as a result of a most interesting experience in international electronic communication. Let me tell you the story. First, though, here is the relevant paragraph, which originally followed right on the heels of the two above:

I have found it amusing to note how much this practice has caught on across the world. Europe, North America, Japan,

Australia. No matter where I travel, homes, especially homes with families, use the refrigerator as the general message board. Wherever I go I find the front of the refrigerator covered with pieces of paper, with notes, with cartoons torn from the newspaper, with schedules of school happenings, parties, family invitations. Pictures, drawings: all the paraphernalia of modern life. The major differences seem to be in styles of magnets.

I believed the paragraph to be true when I wrote it, but during a visit to the University of Colorado, I had dinner at the home of a friend (a German, living in the United States). I told him about my essay on the home magazine kitchen and the worldwide usage of the refrigerator door as a message center. He objected. "No," he said. "In Germany we don't do that."

I couldn't believe him. "But I distinctly remember seeing notes on refrigerator doors," I said. He repeated his objection, and now his wife and children joined with him.

"But look," I said, pointing into his kitchen where the refrigerator door was covered with notices.

"Ah, yes," he replied, "but we are living in America now."

I am fully aware of the distortions that memory can play. After all, I have written books on the topic. One of my standard lines is that the least reliable evidence in a legal trial is apt to come from eyewitnesses. Had I fallen prey to the standard distortion in which our expectations affect the recollection? I decided that I needed more reliable evidence. Fortunately, I take part in a discussion group conducted by electronic mail,* a computer network that links the university and industrial research community all across the world, stretching from Russia to Australia and many of the countries in between. I sent a message describing the section of the home magazine kitchen essay that discussed the refrigerator door as message center. I asked the international readers to tell me whether this was true in their country.

The response was amazing in every way: in the large number of replies (fifty-four messages that filled thirty-seven single-spaced printed pages), in the excitement the topic generated, in the variety of answers, and in how much I learned from the experience.

The responses revealed to me that the manner by which a social group manages informal communications is extremely significant, yet little studied. The home, it turns out, is simply an example of the process. The most important point I learned was the importance of taking a larger view of the social and cultural group. Here, I was scolded (properly) for leaping to a conclusion about message centers without taking into consideration the needs of the living group.

Probably the best response was from a respondent living in Denmark. He pointed out my cultural bias. Yes, I might very well have seen refrigerator doors used as message centers in Europe, but what sort of person would be likely to invite me to their home? It would very likely be professionals who live in modern homes (probably American-style, he suggested), people who travel a lot and who meet Americans, perhaps who have themselves lived in America.

I was neglecting the differences among cultures, he said:

> One must get a knowledge of the common modes of family life in a culture, and they must be described through the members of the family and their activities, who works, who doesn't, what patterns of homecoming and leaving do you find, which family functions are shared (shopping, cooking, eating, watching TV, sleeping, bathing, etc)—all this would be part of the conditions underlying refrigerator door communication. Then you would also have to look at the physical structuring of the home, where is the shared life space, is there an entrance hall, is there a hall way (which are both common in European houses), are there rooms for sleeping, bathing, etc. All these elements congeal into many things, among others they might congeal into a communication board at the telephone in the entrance hall—or in the kitchen, or maybe to your refrigerator door sign board.*

The point is absolutely correct, and I was guilty of missing it. In fact, it really is the point I have argued for in my professional career, so it is rather embarrassing to have neglected it. That is, a design solution has to be consistent with the entire problem to be solved. What are the needs of the group, what are the standard working patterns?

50

You can't just suppose a need or a solution out of context. You have to examine the entire situation.

Just because the American families of my acquaintance need some sort of common message board does not mean that all families need this. Just because many American homes find the refrigerator door to be an ideal location does not mean that other families and cultures will find the same location useful.

In much of the United States today, the kitchen is the social center for the home. There is usually a telephone in the kitchen, and the kitchen serves as a common meeting ground. Earlier in the history of the United States this was not true: As noted in the last chapter, Lillian Gilbreth's kitchen was off-limits to the family except under special circumstances.

In other countries of the world, the kitchen plays different roles in life. Moreover, the large refrigerator-freezer combination is an American concept. I had forgotten my first visit to England where my hosts queried me on the uses of a refrigerator. "Why would you need one?" they asked. "Why so large?" "Oh, I know," I was told, "you Americans like to put ice in your drinks."

Whether messages are needed or not depends upon the needs of the living group. This is apt to vary considerably from country to country, but perhaps even more so among types of groups. The person living alone has different needs for a message center than a person living with others. A family with children has very different needs from one without children, or from unrelated people sharing an apartment. The study of message centers must therefore start out with a study of the interactions of the people, their particular needs and methods of interaction, and then the technologies available to them for leaving messages.

One interesting sidelight that I have not explored fully enough is the kinds of messages people leave for themselves. These self-messages come in a wide variety of forms. A pilot reported that he always pinned a note on the front of his jacket with the words "Close Your Flight Plan" written in mirror-imaged letters so that when he went to the men's room after completing a flight, a look in the mirror would remind him to do this important but often neglected activity. Another person explained that she always wrote short, important notes to herself on the back of her hand: "This way, they are always with me when I need it," she told me.

Technologies vary. The refrigerator door is a convenience because of its large, metal surface, but not all places have refrigerators the same size as those found in North America. Canada, yes. Australia, yes. Large refrigerators are now becoming common in most of Europe and some parts of Japan. That still leaves out a lot of the world. Small refrigerators, as many people pointed out, are usually built-in and too low to serve as a convenient spot to leave messages.

My Danish friend was right—university faculty and professionals are more apt to use refrigerators as message centers than other people, in part because they are more apt to own large refrigerators and perhaps mainly because they are more apt to have lived in the United States for a while where they learned the habit.

But even other American families protested the choice of the refrigerator as the natural message place. In fact, I am now convinced that the need for a common message place—not the use of the refrigerator—is a cultural standard. Almost all my respondents agreed that their families had message centers: the disagreement was where. Of course, my highly educated, technological respondents are all a pretty select group to have continual access to computer networks. So even this need for a message center may be restricted to only a small segment of some societies.

Are refrigerators universally used as a message center? You decide for yourself. Here are the results, listed by country. "Yes" means they are used as message centers; "no" means they are not.

Australia:	1 yes.
Austria:	1 no.
Belgium:	1 yes.
Canada:	2 yes.
Denmark:	1 no, 2 yes. The "no" was qualified as "except in American-style homes."
England:	3 yes. As opposed to Scotland (see below).
Finland:	1 yes. One person commented that "Finns use refrigerator doors much less for posting

	notes than Americans do. Finns tend to use separate bulletin boards in their kitchens."
Germany:	1 yes, 1 no (plus an additional "no" from my colleague who started me on this quest).
Italy:	The respondent did a survey of families in Northern Italy: 4 yes, including the respondent; the rest "no" (the total number asked was "more than 15").
Portugal:	2 no. One person commented that the refrigerator was not in the kitchen.
Scotland:	1 no.
United States:	Everyone who answered said "yes."

Many of the families not using refrigerators to post notices do indeed have other locations. An Italian correspondent points this out:

Families in Northern Italy seem to not use refrigerator doors as a message center. I asked more than 15 medical or psychologist colleagues and the answer has been 'YES I DO' only in 3 cases.

Preferred habit is posting on the kitchen table. Nine persons have also white boards or pin boards always in the kitchen, and append there post cards from friends, commemorative cards and long-lasting messages.

Two General Practitioners told me they have previously suggested their older patients to tape their prescriptions on the refrigerator doors.

Yes, in our family we post on the refrigerator door.

I learned of a marriage counselor, a medical school, and a survey-taker who had all prepared materials that they requested be placed on the refrigerator doors for continual or ready reference.
Alternative message sites included:

The bathroom mirror.
"Inside of the most heavily used entrance door."
The bottom step of the staircase (as both a message center and
    a place to leave things to be taken upstairs).

The floor at the top of the staircase (as a place to leave things
    to be taken downstairs).

Tables were popular: dining room table, kitchen table (three
    times), breakfast nook table, "table in the centre of the
    room."

The front hall (by the only telephone).

Pasted on the hood over the stove.

On the television set (once, my respondent reported, she also
    deliberately turned the set on to attract attention).

On a computer screen.

Shelf above the computer.

Notice boards on the wall, including cork bulletin boards and
    whiteboards.

One correspondent told me of an American high-technology addition: "Of course you have seen the refrigerators with audio tape recorders built into the door to serve the message function?" Nope, I hadn't seen it, but I am keeping my eyes (and ears) open for it.

Some correspondents seemed to think that the very idea of a message center was a reflection upon modern (or even American) life: "You mean you never talk to one another?" I am suspicious of these comments. After all, my family both talks to one another and has a message center. Maybe these comments come from those who have not lived with families. Once there is a large family group, with school, shopping, and work causing varied schedules, with numerous scheduled activities for some or all members of the family, and with children old enough to be receiving telephone calls, there is a need for some sort of social communication center. Certainly people from all countries in my survey responded to the need. Here are some relevant responses.

Portugal:

If you want to leave a message in Portugal, you leave it on the
    bathroom mirror! Or, on the inside of the most heavily used
    entrance door. That's not to say that the Portuguese don't
    like the idea—we've introduced magnets to my wife's family,
    who have begun using them to hang reminders such as when
    to pay bills, etc.

Austria:

Since kitchens are often separate rooms from the rest of the house, the kitchen as such does not become the communication center of the place. It is rather the front hall, where you also find the (usually only) telephone, so messages are more conveniently put up there. Moreover, we usually don't have refrigerators of your size, which means you would have to bend down to read stuff from the door. Another reason lies in the fact that many kitchens are built-in ones, which means they do not have blank doors of metal. It would be impossible to fix magnets on there. And last but not least many people would regard it as a violation of the clean image of a kitchen if you did find pieces of paper and similar put on there. And just another extra thought: Here many women still don't work outside their home, so they still represent the communication center of the family/ house. Therefore paper messages are not really necessary to such an extent. I guess my observations will be true for many Austrian and even southern German households. Meanwhile you will find more families getting larger refrigerators or putting the smaller ones higher up so as to open them more conveniently. I wonder how long it will take until the American message center will get into our culture.

The United States:

Two other places get used for messages: staircase, bottom rung; area around the phone nearest the kitchen if phone is not next to the refrigerator.

We have an odd kitchen because one of the walls is all glass. So we bought an additional cutting surface and placed it between eating area and work area. It has also turned into a message center. Obviously the staircase only works if one lives in a 2-story house and there is a guarantee that the paper will stand out against a background. I agree that it is crazy that this function has not been scripted into the design planning stage.

Many correspondents pointed out that although the refrigerator door, bulletin board, or other central posting facilities were used by families, they often could not be characterized as "message centers." What do people put on the boards? Schedules, newspaper clippings, cartoons, artwork (usually by children), and so on. Some items stay for years, so their original purpose has long since disappeared. English respondents liked to comment on the prevalence of children's magnetic alphabets, with letters stuck on the refrigerators, sometimes to spell out words. One person even suggested that the contents could be used as a personality assessment of their creators. I must admit wondering at some of the things I have seen displayed. This is especially true as I visit universities across the world and see the various newspaper clippings and cartoons that faculty members and graduate students post on the outside of their office doors for passersby to read. These clippings turn the stern, official-looking professor into a person, with opinions and feelings like everyone else. They often reveal a side of the person that would otherwise go unnoticed to the student or casual visitor.

Finally, many correspondents commented on the cultural issues involved in these studies. One British correspondent who has studied communication patterns of everyday people (that is, nonacademics) stated:

> It was very common for people to have a set place in the house where they would leave messages and in a paper we actually give the example of messages stuck to the fridge with magnets. I am impressed with how most households are organised for reasonably efficient communication, but that there are several quite different ways of doing it.

In doing this study I soon found myself captured by the interesting range of phenomena that were discussed, interested enough to start taking the problem seriously. I am now considering studying the informal means that we use to provide message and communication centers among family members and colleagues. I am also interested in the ways that new technologies might aid in this practice, aid without interfering.

One correspondent mentioned that a researcher at Apple Computer had developed a program informally called, appropriately enough, "Fridge Door." This program was intended to make it easy for members of a working group to post messages to one another, messages that appeared on the face of the computer display and looked something like refrigerator notes.

By coincidence, shortly after receiving that note I visited Apple's San Francisco Multimedia Research Laboratory where the work had been done. I talked with the program's developer. Yes, it was intended to be used as a message center, with displays located throughout the laboratory. In one example that he showed me, the Apple team had been working with a public school, and the display showed pictures of teachers and classrooms. The notes were actually audio/video clips. If you moved the mouse cursor to a note and clicked, you got a short video segment and an audio description. Lab workers added audio comments to the note. In this way, people who did not visit the school could get a quick overview of the people there, the projects underway, and other lab members' comments.

As a display and communication device, Fridge Door seems to be very effective. However, the task of preparing the video and audio clips for presentation is difficult enough that the Fridge Door program is not as widely used as the lab group would like it to be. But what a wonderful idea! Capitalize on the universality of message boards, expanding them to allow for real photographs and videotapes, along with sound, as well as a way of letting each viewer add further comments. Moreover, a bulletin board that can be located at many different locations, all with the same information. A worthwhile study indeed.

Postscript

When my electronic survey was over, I posted a summary of my findings to the electronic mailing lists (the summary being, essentially, the first draft of this chapter). This, of course, led to more responses. Best of all was the following note (from an American living in Canada):

On Sunday night we acquired an "exchange student," a daughter of a colleague in Bonn (Germany) who is visiting us till October. When your summary came in tonight I went immediately out to the other room to ask her what was on her family's refrigerator.

"It's in a cupboard," she said. "Nothing is on it. It's neater that way."

They have, she told me, a bulletin board in the kitchen where notes are pinned (the kitchen, of course). My daughter reminded me that we ourselves don't leave notes on the refrigerator; we have a blackboard in the kitchen, which is used incessantly, for notes and for various kinds of jokes. But she also reminded me of the William Carlos Williams poem, which is clearly a "refrigerator note":

> **This Is Just to Say**
> I have eaten
> the plums
> that were in
> the icebox
>
> and which
> you were probably
> saving
> for breakfast
>
> Forgive me
> they were delicious
> so sweet
> and so cold

5

High-Technology Gadgets

I STUDY cognitive artifacts, these artificial things that enhance our thinking. Ever since I started in this business, I discovered an interesting side benefit: It became my solemn responsibility, my sworn duty as a scientist and public-minded citizen, to try out each new device.

It isn't always pleasant. It's hard work, this business of trying out new technological advances. Many of them first make their way into entertainment systems, so I am forced to endure theaters, game arcades, amusement parks. I spend hours in such places, watching the new technologies at work. It's hard work, I tell you.

In my home, I feel compelled to buy one of everything. It's fun, especially when some of the gadgets are games or entertainment systems. The one with the most promise is "virtual reality," a system in which you slip on goggles that contain a color computer-generated television image for each eye, which makes you feel that you are in another world, a "virtual," artificial world. You wear an instrumented glove or even a suit that allows the computer to know your every move, every change in position. The scene in front of you changes as you move. You can fly through the air, wade through water, explore imaginary places, or for that matter, you can explore real places that are too difficult or dangerous to get to—for example, Mars, or the inside of the body, or an architect's plans for a house not yet built.

Alas, the promise of the laboratories is not yet in existence for the home. The closest so far is a spin-off from the research, a glove. You know, something that goes over the hand. In video games, there has to be some way of telling the computer what part of the visual image is of interest. Usually this is done by some sort of joystick, a rather indirect way of interacting with the world. With a glove, you simply point and the glove itself is instrumented to determine where it is being pointed, which fingers are bent, and which are wiggling. After my first experience using a laboratory prototype system for generating "virtual reality," I rushed out and bought a "game glove" to try out on my son's video games. Imagine a video game boxing match where you actually punch at your opponent, or a traverse through a computer-generated world where you point at where you wish to go. Not earth-shattering, but interesting.

The market for future virtual reality systems seems large, especially in entertainment, although it has more practical uses as well. Imagine putting on a helmet that monitors just where the head is pointing, where the eyes are looking, with a small video screen displaying an image for each eye. Imagine putting on a body suit so fully instrumented that every move and wiggle of the body can be sensed by the computer. Imagine being able to explore a proposed home or building, seeing it in three dimensions in front of you, being able to walk through the proposed hallways and rooms. Imagine the same system used for exploration: See and experience the planets, travel through the game reserves of deepest Africa, go up the tributaries of the Amazon. The business, scientific, and educational opportunities seem endless. Purveyors of entertainment systems are eagerly awaiting the day when this becomes practical at an affordable price.

Will it work? Who knows. Will the price be reasonable? Who knows. Will users be willing to encumber themselves with suits, gloves, and helmets? Who knows. That is one of the reasons I thought I would try the simple game glove.

Surprise: The several-thousand-dollar glove I used in the research laboratory version of virtual reality seemed to work better than the $80 version I bought at the local toy store. The home glove now sits unused, in a corner, in a pile of other unused, no longer enticing gadgets. Maybe it was too crude, the relatively low price required for a household item not allowing for sufficiently precise electronics and sensors. Maybe the system wasn't properly integrated so that the full benefits could not be tested. Maybe the idea itself was wrong: Perhaps the game glove is simply a bad idea. Which is it?

A few years ago, the video-game maker Nintendo offered what I thought a brilliant innovation in games: a floor pad. This was a simple flat pad that you placed on the floor and stood on, the pad sensing the location of your feet. I imagined races or exercise programs in which I, as user of the system, could be pitted against video runners on the screen. The race would begin and I would start running in place, on the pad. The pad could tell how fast I was going by the timing between footsteps. Runners would pass me unless I increased my pace to whatever the program deemed most acceptable for my exercise program. Imagine that: computer runners who went at just the speed I was supposed to go to get my daily exercise, meaning that if I kept up with them it would be neither too fast nor too slow.

I could even imagine hurdles that I must jump over: tree branches, streams, whatever. The pad could sense the elapsed time between my feet leaving the ground and returning again. Couple the timing with how fast I had been running and the system could make a good estimate of how far and high I had jumped. What a wonderful way to exercise! I never got a chance to try it. The device never caught on and disappeared from the stores before I got a chance to buy and test it.

A number of years ago a company offered a fancy expensive gadget with a similar thrust: to make exercise bicycling more enjoyable. You sit on the exercise bicycle and pedal away while watching a television screen that shows you traveling through some exotic location of the earth. Pedal quickly and the world passes by quickly. Pedal slowly and it goes by slowly. Come to a fork in the road and you can decide which way to go by turning the handlebars of the bicycle. The price was outrageously high and someone once told me that not a single one had even been sold. Ahead of its time? Or badly done? Or just a bad idea? It would be interesting to know which.

The same theme has been taken over by many makers of exercise machines. In fact, I was once called by a magazine journalist: "We are doing a story on those new high-tech exercise machines," he said, "and we wondered if you could comment on the effectiveness of their strategy."

Well, my solemn responsibility to the world of gadgets being what it is, I had already looked these machines over. I had discussed them in my class on design and had even included a question about them on a midterm examination. These exercise machines follow a common theme. Most display the time, your pulse rate, how fast you are going, and sometimes an estimate of how many calories are being expended (a number that turns out to be woefully inaccurate). Some allow you to program the course, to have periods of hills interspersed with level ground so that the running or bicycle riding vary in difficulty. Rowing machines tried displaying cartoon-like competitors on an accompanying television screen, so you have to adjust your rate of rowing to match that of the competition, much as I suggested could be done with the Nintendo sensing pad. All in all, there are a variety of techniques, but all of this same general character.

Once again I decided to try out these ideas. I set off for the exercise rooms at the university. I even stuck my head into a number of commercial exercise centers to watch the machines and the participants. I brought up the topic in class again. The students were quite knowledgeable. Many exercised regularly, and among them, they had tried out almost every example of the new high-tech machines, either in their homes or, more frequently, in the university weight and training rooms, and in health clubs.

Exercise is fundamentally a dull and unpleasant activity. The slogan "no pain, no gain" may be overrated, but the general philosophy it expresses is widely believed, and it certainly does not seem calculated to inspire people to enjoy the activity. So, the question is, how can you get people to do unpleasant, dull activities? Even though exercisers know that the exercise is good for them and they probably even enjoy portions of it, and even though most do enjoy the feeling afterward, exercise is still distasteful and difficult to do, for most people. The wonder is that so many people keep at it.

My suspicion is that the hardest part of exercising is probably that of getting started in the first place. That is why social structure is so good: Make an appointment with someone to exercise together at a specific time, and you both show up because the social structures force you to.

Yet another problem with exercise is sticking to it for the assigned period, and at the correct rate—neither too slow (or else it does no good) nor too fast (or else you can damage yourself), and with the proper warm-up and cool-down periods. Finally, there is the real problem of continuing the exercise sessions for a lifetime. The drop-out rate among exercisers is enormous. Many people start a sensible exercise program, but the number that are still at it several years later is low.

The new high-technology exercisers seem mainly directed at the problem of doing the exercise at the proper pace. Some try to tackle the motivation problem of keeping at it month after month, year after year by varying the exercise routine from day to day. The vicarious traveler—that bicycle-video system—tried to solve these problems by making the exercise intrinsically interesting and educational. The Nintendo exercise pad—or at least my imaginary dream of how it could be used—did the same by continually changing the game being done.

The experience of the people I have interviewed is that all of these attempts fail after a few months. The two hardest problems, they state, are getting started in the first place and then keeping at it. Looking forward to a little television screen with clever messages or pretty pictures just doesn't do anything for them, they told me.

What if we had real virtual reality? What if instead of watching your competition on little television screens, displayed in poor resolution through computer graphics, you felt as if you were really there, really on safari in the jungles of Africa, except that you were minus the preying bugs and in an air-conditioned climate? Or what if the racing companions were always there, always scheming to beat you, remembering what had happened on previous occasions? Would this make a difference? Would full virtual reality, complete with three-dimensional vision and sound, transform the dreary task of exercise into an educational, inspirational experience? Maybe even one you looked forward to each day?

I myself have discovered that exercise is far more rewarding when there is an attention-absorbing activity. In my case, I have found a relatively low-technology solution. Many people now go on their daily run with little tape recorders or radios, listening to music or news as they run. I have discovered what to me is an even better solution: language tapes. Now, as I run, I listen to language tutorials. First it was Spanish—sixty tapes' worth. Now I am starting French. Language tapes have the virtue that they are completely absorbing. You have to be an active listener, attending carefully to the sounds and then attempting to recreate them yourself. Radio shows and music are passive: Just listening to them while running isn't sufficiently distracting. Moreover, there is nothing special to look forward to for the next run, no real reason to want to listen to tomorrow's radio show or music.

My language tapes force an active process of listening and responding that absorbs my attention so completely that I am sometimes surprised to discover that my run is over. For years I failed to take language learning seriously, always putting it off to some other time. This seems a perfect solution. Put two tasks that you should do but wouldn't otherwise do together and let each motivate the other.

Other attention-absorbing materials might work equally well. One possibility is recorded books, audiotapes of published writings, often recorded by the author. Would these capture the attention in a satisfactory way? According to my mini-theory of the role of attention in exercise, they would succeed if the listener got actively involved in the story or material. In fact, according to this theory, almost anything will work if it requires active mental participation and taps into the listener's special interests. Thus, listening to a sports match over the radio will qualify to a sports fan, for the announcer's verbal descriptions are actively restructured by the listener into a visual/spatial mental image of the events.

My experience with the language tapes and my mini-theory that active participation and involvement can be captured by a variety of materials bolsters my faith in all these gadgets. Although a portable tape recorder is low-technology if compared to virtual reality systems and some of the expensive exercise machines, it actually is sufficiently advanced that even a few years ago, it would not have been possible to find these rugged, lightweight tape recorders. It is also only recently that language courses have been developed for listeners who are not simultaneously required to read from a book. I still need to use a book to learn to read in my foreign language, of course, but I can't use a book while running along my narrow, sometimes close to the cliff-edge running path. Fortunately, language can be learned without books, as the experience of every two-year-old proves.

Language tapes are not for everyone. Nevertheless, this and related experiences lead me to conclude that the new technologies have real promise, even though I am more often disappointed than not. There are real benefits to some of these new technologies, but, as yet, they are not well understood, not well explored. In part this is because the correct environments haven't been set up. In part it is because the technology isn't completely here. So far, the technology is too expensive and too fragile for real use, and the simple versions that do filter down to the home video game or exercise market are too crude to capture the full potentials of the methods.

Why do these devices always seem so exciting when I read about them, or test them briefly in some laboratory demonstration, or try them out at a store? And why do they fail so miserably at their promised returns when in actual use? One reason is that imagination is far more powerful than reality, far more tolerant of flaws and defects. Another is that short tests do not reveal the flaws of design, the difficult and awkward parts of regular use. Somehow, in the demonstration, it all seems to work smoothly. It is only when the device gets tested in the cruel reality of the living room that the defects appear. Worse, the defects dominate.

Alas, each new device is incompatible with each other device. Each does something good, at the cost of something new to be learned, something awkward and clumsy to be avoided, some disastrous side effect to be guarded against, and yet something else to lose, to trip over, and to break. But the real problem with many of these fancy new devices is that they are badly conceived, developed solely with the goal of using the technology. They ignore completely the human side, the needs and abilities of the people who will presumably use the devices.

The computer is often held up to ridicule as the ultimate example of poor design. Let me disagree. Many computer manufacturers now provide important safeguards for their users. Thus, computers have reached the stage of reliability where we seldom need worry about losing information. Not only do the application programs guard against loss and error, but there are also reliable means of backing up, of protecting ourselves from mishap. These gains did not come without cost, and they are still not universally available. Nonetheless, many companies within the personal computer industry have learned the importance of taking the user into consideration, of aiding the learning process and helping guard against error. Alas, this lesson does not seem to have been learned outside of that industry. The designers of these new devices seem determined to learn the hard way, repeating each and every error that for so long plagued our computer systems. Each new industry seems unwilling to learn from what has gone before.

Consider my high-tech pocket electronic address book, yet another new gadget that I purchased, lured by its advertised benefits, captured by my own desire to try each new technology. Small and compact, it contains hundreds of names, addresses, and phone numbers, and my daily schedule. I carry it with me at all times. But it is an unruly monster. It eats appointments, chews up notes, and mangles phone numbers. It suffers from what I have come to recognize as the Japanese Engineer's Design Syndrome:* tiny little buttons, each labeled with tiny dark letters on a black background. As for the design of the functions? Reminds me of the computer programs of the 1960s: Push the wrong button and lose all your work; forget to push the right button and lose all your work. Lose if you do, lose if you don't. Critical functions are called up through obscure keypresses. Unimportant, never-used functions are operated by big, clearly labeled keys.

The pocket electronic address book is a perfect example of my continual love-hate relationships with today's technology. It provides valuable functionality that at times seems an essential part of life, but at a horrible cost, a cost measured in time to learn the over-complex operations, time to recover from errors, and time spent worrying that some error or mishap will cause all the important information that it records to disappear to an electronic graveyard, never to be recovered.

The list of other failures seems endless. Perhaps the worst problem, however, is the amazing proliferation of gadgets, each one incompatible with the next one. I could learn to use any single device, no matter how complex, no matter how bizarre. The real problem is that each one seems to differ from all the others—not just differ but actively conflict. Each device conspires against the others. Each requires electric power, either batteries or wall current. But if it requires wall current, it is apt to be through a wall transformer, a little black box that plugs into the outlet, heats up, hums loudly, and physically protrudes just far enough to prevent the adjacent outlet from being used. "Ha, ha," the device seems to say, "I got here first and nobody

66

else can be here too." Four gadgets and you have used up eight outlets, which is all the average room has. If these devices show such antisocial behavior when you plug them in, even before you have turned them on, what can you look forward to once they start operating?

Each gadget breeds more gadgets. So they use up your electric outlets. The solution, of course, is yet another set of gadgets: outlet strips. In my home, we solved the outlet problem by buying what seems to be dozens of outlet strips. Outlet strips used to be simple. No more. Some have built-in lights and switches. Some have circuit breakers. Some have electronic "solid-state" circuits to protect against power surges and noise. We even have one that can be turned on and off remotely. And one is so intelligent that it sits by my computer sensing its state so that the instant I turn the computer on or off it automatically does the same to the five other things that have to be powered on and off with the computer. That one came with its own set of programs on a floppy diskette. What a sad commentary on the state of household gadgets: an outlet strip that comes with a floppy diskette.

My home computer contains all my schedules, my addresses, my financial records, and most important of all, my writing and ideas. So I had to buy a special outlet to protect it against harmful electric power. And a special "backup" storage system so that I can duplicate all my records each night onto a separate storage cartridge, in case the computer fails. And a fireproof safe in which to place the backup cartridges.

My television set is covered with boxes, switches, and wires. Two video recorders, several video games, and a switch because we need to plug more signals into the television set than it was built to accept. Many people find that the video recorder, the television set, and the television cable channels present such a complex mix of technologies that they have to purchase yet another device that simplifies the operation of the video recorder. Today's gadgets are so complicated that we need to buy gadgets that have no function except to help us use the other gadgets.

A company I consulted for graciously lent me a $100 educational program that allows its computer to control a video laser disk in a sophisticated manner, showing wonderful examples of educational materials on subjects ranging from physics to geography. But to use this loan would have cost me about $800 because I would have had to purchase a laser disc player that was capable of being controlled by a computer, and then move the computer into the television room. This one program, enticing though it seemed at first, would have meant yet more money spent, more boxes by the television set, and probably yet another set of switches for both the computer and the television set to accommodate yet one more set of devices.

Where will it all end?

On top of these problems, the devices also conspire against intelligence. Each gadget works differently from the others, even when the others do essentially the same task. If one device has a numerical keyboard with the numbers 1, 2, and 3 on its top row (just like the telephone), another will have 7, 8, and 9 along the top (just like a calculator). And a third will have the numbers in two long horizontal rows.

We have two remote control units, both made by Mitsubishi, both capable of controlling a video recorder and a television set (one came with the television set, the other with the recorder). The two are functionally equivalent with almost the same set of buttons and controls. Yet they are organized in a way guaranteed to infuriate: The controls for one are located in the opposite positions from the controls for the other:

The Television Controller			The Video Recorder Controller		
CHANNEL		VOLUME	VOLUME		CHANNEL
UP	MUTE	UP	UP	Q.V.	UP
DOWN	Q.V.	DOWN	DOWN	MUTE	DOWN

Volume control on the right for television, on the left for recorder. Channel selection on the left for television, on the right for recorder. Mute button above "Q.V." for one, below for the other. (What docs "Q.V." stand for? "Quick View." Is it useful? In principle, yes. In practice, no, but that's another long story.) The major consistency is that they are both designed true to the fashion of the Japanese Engineer's Design Syndrome: tiny little buttons, each labeled with tiny dark letters on a black background, in this case on controls that are meant to be used in the dark, where it is difficult to remember which one of the pair I have just picked up. Why are they designed in such a mindless fashion? Why can't one manufacturer be consistent with its own designs?

And then there are the conflicting operating instructions. Some things have to be turned off before new cartridges can be loaded; some have to be turned on first. Some have warnings not to do X, others have warnings that you must do X. Is it any wonder that my family and I have mixed feelings about these new technologies? And the worst problem of all has not been mentioned: They require continual maintenance, continual carekeeping.

Consider the problem of clocks. How many clocks do you have in your house? In my house, we have—are you ready?—twenty-seven. I don't believe it either, but this is a topic I explore in Chapter 14 ("Hofstadter's Law"). Some clocks are hidden away in timers, such as the clock inside the refrigerator that controls the defrost cycle, or the clocks that control the outdoor lighting. Seven are wristwatches. (I have no idea why a family of three requires seven watches.) In my bedroom, our clock radio, of course, has a clock, the digital scale has a clock, and there is still an old alarm clock, now replaced by the clock radio. Even the telephone has a clock. Five more are accounted for by our two videocassette recorders, the two television sets, and the audio set.

Why all these clocks? Why does my scale have to have a clock? I have even seen clocks on musical keyboards. Maybe these clocks give the scale, telephone, and portable radio something to do when they aren't showing weight, phone numbers, or radio stations. I guess. Maybe watches are like clothes hangers, they secretly reproduce in closets and drawers when no one is watching.

The real problem is not the number of clocks, it is the need to reset them every time the power fails. Or when their batteries die. Why do I have to go around the house resetting all the clocks after every power failure? Or twice a year when we go between daylight savings time and regular time, setting each clock ahead or behind one hour, after first spending untold time trying to remember which way it should be. I can't even find the clocks, let alone reset them. And some of the digital clocks can't be reset without first finding the instruction manual: Remembering how to reset the clocks becomes the task of the month for the family problem hour.

There is a simple solution, if only I could convince the manufacturers to go for it. Suppose each house had one master clock. It should be designed so as to be easy to set, following all the best traditions of user-centered design. This one clock could be powered by anything you like, but the critical thing is that all other clocks in the house would get their time signals from it. The master clock could do this by sending a special, digitally coded signal over the home's power lines. All the other clocks in the house would monitor the power line and display the time sent out by the master clock. Some of these other clocks wouldn't even have to be clocks—they could simply be displays. If there were a power failure, or at daylight savings time, only that one master clock need be reset—all the others would gracefully follow along. How about that?

I could go on to talk about graceful ways of coupling magnetic signals to wristwatches and battery-operated devices. Or how to avoid clashes with other devices that send their signals over the power line too. Or about the problems of error when all your eggs are in one basket—that is, when all the clocks in a house are controlled by one source. Or about how even the home's master clock could be replaced by a region-wide time signal (international standard time signals are already broadcast on special radio frequencies by the Bureau of Standards). But I will resist the temptation.

And what about all the other daily chores required by our gadgets?

Replacing batteries
Maintaining everything (oiling, watering, cleaning, checking, and tightening; repairing belts, checking screws and pads, replacing filters and batteries)
Synchronizing address books and calendars with spouse, family, secretary, home, office, friends, computer, noncomputer
Backing up computer files

How might we eliminate these chores? Come on, inventors of the world, there are riches awaiting those who can simplify our lives: Just simplify the use and maintenance of all the gadgets that are supposed to simplify our lives. (And don't forget to make your solution simple to use, simple to maintain, simple to simplify.)

At first my family joined me in my pursuit of gadgets. But after a while, the problems seemed to multiply faster than the benefits. There are times when we'd just as soon throw all our high-technology gadgets away and live in a small tent on a secluded island. My family has threatened: "No more gadgets. One more and we leave." I sympathize, and in fact, if they leave, I will join them. We fantasize about getting away from all of this. Just us and nature. A nice tent. Some comfortable chairs. Some good lights. A good audio system and a pile of compact disc recordings. Some books. Notepaper. The daily newspaper. Our favorite monthly magazines. A computer to jot down ideas. Maybe a coffee grinder and coffee maker. A fan and a heater for warm or chilly nights. Just the family and nature.

6

The Teddy

THE delights of having information ever-present are amazingly seductive. Wouldn't it be nice to have a personal assistant, small and unobtrusive, that could remember the details of life for us, so that we could always have them available on demand? It would take care of the daily trivia of life, things like telephone, passport and driver's license numbers, as well as the important things:

> "What was the name of that wonderful restaurant we had dinner at two years ago?"
>
> "How late does the library stay open during the summer? Do I have time to get there?"
>
> "Is today Michael's birthday? I forgot all about it. Quick, I'll buy him something. Umm, what size does he wear?"
>
> "Remember good old what's-his-name, you know, the person we had dinner with after that party at what's-her-name's house?"

Or even new information:

> "I have a free evening. What's happening in town tonight? Any tickets available?"

We are particularly bad at remembering, especially when it comes to details. In fact, we aren't very good at anything that requires great precision and accuracy. We remember the major experiences of life, but less accurately than we might like to believe. The details fade quickly. Our strengths lie in other areas: in aesthetics, beauty, humor, and imagination. We are excellent at doing meaningful things, at understanding, at leaps of creativity. In fact, I believe that the very brain mechanisms that make us so good at creativity and aesthetics result in poor capability for the memory of details.

You would think that if we were good at one thing and bad at another, we would structure the world to emphasize the things we excel at and minimize those we aren't so good at. Nope. We did create a world that appreciates creativity and beauty, humor and insight, but we also managed to create one that requires precision and accuracy. In today's world of technology, of machines, clocks, and computers, details are essential. Arbitrary numbers rule our lives: phone numbers, mail codes, street addresses, government identification numbers, driver's license numbers, bank account numbers, passport numbers. Somebody always seems to be wanting us to provide a number, some new form seems always to want precisely those details that we cannot remember. The problem is that the same machines that we invented to aid us also demand precise, accurate information from us. As a result, we often end up the slaves of the very machines we have invented to serve us, slaves to their relentless demand for precision, accuracy, and continual supervision.

But, thinks the ever-optimistic technologist, the problems created by machines can also be overcome by machines. Suppose we could make machines to aid us in providing these details, the minutiae of life. Suppose we could give every one of us a little personal assistant, a helper that we would carry with us everywhere, that would continually give us access to the information we need to go about our lives quietly and efficiently.

You can, I am sure, make up your own list of things you wish you could have remembered at the time you needed them. Suppose you could somehow record them readily and easily with a device that was ever present, always at your service, but small enough not to be a bother. Suppose we could invent the perfect mixture of artifact and human, each serving the other well.

We humans, as I have said, are good at the creative things, no so good with the details. Computers are just the opposite. Why not marry the two properties: Create a computer-like device that is small and portable, and unfailingly accurate, retentive, and precise. Imagine, if you will, a science-fiction scenario. You may view it with delight or horror, but either way you look at it, it is a reasonably plausible idea. So bear with me for a while.

The Teddy

Let's assume we've reached the time when the power of information technology has increased enormously, with the whole country—nay, the whole world—wired so that anyone anywhere can connect to the huge communication network. As a result, society has evolved to the point where everyone always carries a portable computer with them, except that it is thought of not as a computer but as a personal, confidential assistant.

Each of us would have our own portable device, all the time. In fact, by starting out with a personal assistant at a very early age—from two or perhaps three years of age—it could help us learn to read and write, draw and sing, spell. Because the devices would be handed out early in life, the version for young children would be soft and cuddly like a Teddy bear—hence the name "Teddy."

By starting so young, the Teddy could store within itself all the information and experiences of a lifetime. We would become quite intimate with our Teddys. They would know all about us, while also giving us complete access to the world's databases of knowledge. I assume that by the time such devices are possible the speech recognition problem will certainly be licked, so we could communicate with Teddy by talking. We talk to it, it talks to us.

Teddys would be with us for our entire lives. They would change in shape and form to match our growing sophistication and interests, but each time we got a new model Teddy, the information from the earlier version would be transferred to the new. As a result, Teddy would always retain a complete record of all of our personal experiences and knowledge for an entire lifetime even as it changed in physical form.

Eventually, as we came to rely upon our Teddys, we would reach a point where we would be disoriented without them. After all, with a Teddy, we would never be alone. We could always talk to Teddy. It would never desert us. It could be programmed to give reassurances, to follow progress on a task, and to make appropriate suggestions. It could serve as a continual reminder of names and dates: time to do exercise, to phone home, buy gifts, etc.

With a permanent Teddy, memorization would no longer be needed: We would just tell Teddy all we want to remember and then have it do the remembering. Of course, with all our thoughts, ideas, and memories contained within an artificial device, we would have to make sure Teddy was always with us. Without Teddy, we could no longer function.

People today can no longer memorize lengthy poems and orations as in times past, for the art of memorization has all but disappeared as writing and audio and video recorders have proliferated. Skill at arithmetic deteriorates as we rely more and more on calculators. Teddy would accelerate this reliance upon technology. The technological solution is to ensure that we would never be without our Teddys: We would attach them to our bodies, much as today we attach our watches to our bodies. Maybe our Teddys could even be surgically implanted inside our bodies so we could never misplace them. They would interact by voice, a small microphone surgically implanted near the throat, a small loudspeaker or earphone surgically implanted in the head.

Implications

The Teddy is the stuff of science fiction. It is easy to get carried away with the theme, to imagine results either wondrous or horrid. The problem is that it really isn't fiction; it is very likely to take place.

Can you imagine the future? People walking around with little lights glowing in their foreheads, indicating whether their Teddys were on or not. People going around apparently mumbling to themselves but actually conversing with their Teddys. New forms of mental disturbances could appear: Is that voice talking to you at night a manifestation of schizophrenia, or simply an overactive Teddy? At parties, some people would cuddle up to their Teddys and ignore other people. Others might be openly defiant, making a show of turning off their Teddy, defying it to complain, insisting that anyone they talked with do the same. Parents might have to instruct their children on proper behavior: "It isn't polite to talk with Teddy at the dinner table." All this from a device intended to help us remember numbers and dates.

Let's imagine a debate about the necessity of being able to turn a Teddy off. After all, if Teddy records everything we say to it, it could also record everything said to us. Now imagine this scenario: Congress passes a law saying that to protect civil liberties, all Teddys must be able to be turned off, with some indicator so that others would know whether or not a person's Teddy was listening. Then again, let's imagine a different scenario, one in which failure to have a Teddy on at critical moments leads to unfortunate incidents. In this scenario, Congress might pass a law forbidding it to be possible to turn Teddys off.

Many people seem to have a need for artificial memories or for a quiet, nonjudgmental confidant. The written personal diary serves somewhat the same purpose. Diaries have problems, however. They are hard work to create: Many more diaries are started than are continued. Worse, because they often contain private thoughts and information, they cannot be left where others might gain access to them.

The Teddy would solve all these problems by making it easier to record thoughts, using voice, keyboard, handwriting, and perhaps even photographic devices. It would be small—as small as you wish to imagine—and private, with the most advanced cryptographic techniques rendering the information useless for anyone but its owner (although I can imagine that the legal arguments about when it would be proper to force an owner to reveal the contents could keep thousands of lawyers occupied for many years).

How likely is this? Very. Even today we see people carrying around their primitive Teddys. Pocket appointment books, address books, notepads. Some are relatively big and bulky, notebook-sized, but too big for the pocket. More and more are electronic, tiny little devices made for entering names, phone numbers, appointments, and even memos. The one I am experimenting with is small enough to fit into most of my pockets (except then I can't sit down), is much smaller than the address books and calendar I had been using to contain the same information, and even comes with an adapter that makes it capable of sending faxes previously typed on its tiny keyboard.

Portable computers already can fit in a shirt pocket. Voice recognition systems are primitive, but they do exist. Each year these new tiny artifacts become more powerful. The progress may appear slow to us, whose lives span less than a century, but for the pace of human history, where a hundred years is barely noticeable, the pace is rapid indeed.

I am intrigued by the fact that we like some artifacts, such as the wristwatch, so much that we strap them to our bodies. We need a watch only in a society that synchronizes all activities by reference to time. Do we need entertainment so much that we must strap the devices that provide it to our bodies? There are some who seem never to be without their plugged-in earphones. And soon, they will never be without their plugged-in eye-set, glued to a video image of this or that, whether fact, fiction, or fantasy. Readers too fall prey to this disease. Why always with book in hand, book to the eyes? Reading while waiting, reading while traveling, reading at the dinner table even with others present.

Why is it that some people never wish to be alone, never allow themselves a quiet, reflective moment? The popularity of portable entertainment systems seems overwhelming. Books, magazines, comics. Wearable radios and television sets. Why aren't some people comfortable with their own thoughts, alone, in privacy?

One reason might be that isolation breeds paranoia. It is easy, when alone in the midst of night, to think of some past or future event, imagining as it was or will be, but going beyond what was there. One imagines the comments of colleagues, and the comments to those comments, which, of course, were never even thought, let alone said. One imagines events taking their own courses, events that never happened. Wondrous and horrid things take place in the solitary mind. It is too easy to develop a subconscious fantasy in which you are the butt of all others. A fearful spectacle in which you think the worst of all. Supportive colleagues withdraw their support. Casual enemies become dreaded villainy. The slight mishap of yesterday becomes a major disaster of tomorrow.

Or the other way around, of course. The unaided mind can just as easily fly into euphoria. A slight advance of yesterday makes you the hero of tomorrow. Slightly supportive friends become heroic champions, disciples even. You wander through your own fantasy land as superhero, supergod. Whichever way it is, the unaided mind

77

has lost all touch with reality. The quick cure is once again to encounter reality, to substitute the real world for that fantasized one.

What has all this to do with the Teddy? Actually, it seems an argument for a constant companion, an argument against isolation. That depends upon the nature of Teddy.

In everyday life, positive means good, negative bad. However, in systems with feedback, it is often the reverse. Negative is stabilizing, calming. Positive is encouraging, reckless. Negative feedback is what lets you drive a car at a steady pace, keep a room's temperature at a constant value, control an airplane to maintain its altitude, speed, and heading. When you set an automobile's cruise control at some fixed speed, it establishes a negative feedback loop. Whenever the behavior of the car deviates from the desired setting, the feedback loop corrects by making the car do the negative action, which translates into the opposite. If the car is starting to go too fast, the controller makes it go slower. If the car is starting to go too slow, the controller makes it go faster.

If the controller worked by positive feedback, it would be a disaster. As long as the car stayed at its assigned pace everything would work fine. This is what we call equilibrium. But whereas the negative feedback circuit yields stable equilibrium with every deviation automatically recovered, the positive feedback one is an unstable equilibrium, where every deviation leads to a greater amount of deviation.

Suppose the car controlled by a positive feedback loop starts to slow down. The positive feedback says to do even more of the same—to go even slower. Having gone even slower, the positive feedback circuit senses the deceleration and says to go yet slower. Eventually the car will stop.

What if the car goes slightly above its set speed, slightly too fast? Well, the positive feedback circuit will make it go even faster. And that faster speed, being detected by the positive feedback circuit, will then lead to an even faster one, until finally the automobile goes out of control.

Similarly, dreams and fantasies can feed upon themselves through a positive feedback loop, each turn reinforcing the hopes or the fears of the previous, until the dreamer has distorted reality into a foreign existence, far removed from what is possible or even likely. Positive feedback can lead the dreamer to lose touch with reality.

How should our Teddys respond to our dreams and fantasies? Should they follow the route of positive feedback, always being supportive, always encouraging? Or should they follow negative feedback, always being critical, corrective?

Consider the supportive Teddy. If we tell it our fears and it confirms them, we are driven to even worse fears. If we tell Teddy our most wonderful fantasies and it confirms them, we will be removed further from reality. A positive feedback Teddy could be harmful. But a critical Teddy would not help either. If every time we dreamed of something wonderful, Teddy voiced its doubts, well, it would be like living with a nagging parent. It might work better on the downside: Every time we voice doubts and suspicions, we are met with a reassuring response. Except that some doubts are necessary and useful. Just as some wild fantasies keep us excited, dreaming of and even accomplishing events that once were only dreams, doubts keep us realistic, forcing us to consider alternative courses of action, causing us to plan for the unexpected.

To get the correct balance of support and criticism will be difficult. Many people have never managed to find it in their relationships with others, or even themselves. How can we expect the anonymous programmers and engineers of the future who will design the Teddy to do better?

There is another argument against the continuous presence of a companion. If we are never alone, never in quiet, when would we think? There is a difference between the mental stimulation created by signals from the outside and that which is self-created. In the first case the mind can be passive, simply responding to and enjoying whatever it is offered. The scientific term for this mode of operation is data-driven processing, where all that goes on is driven by the arrival of sensory data. It is a necessary part of brain functioning, but it is externally driven, which runs the danger of also meaning externally controlled. In the other case the mind has to drive itself, to develop and invent new concepts and thoughts. Now the mind is active, creating for itself the images and thoughts that will occupy it. The scientific term for this mode of operation is conceptually driven processing. This is the inventive, creative part of life. This is the mode in which new thoughts and ideas can arise. Normal processing requires both modes of operation. Excesses of one mode over the other have different implications.

Excessive stimulation—too much data-driven processing—leads to an externally driven existence, passively accepting the guidance and information from others. A complete lack of stimulation—too much conceptually driven processing—can lead a person away from reality. The extremes of this situation are the drug-induced experiences and hallucinations created by mind-altering drugs, including alcohol. Or perhaps the extreme paranoia or euphoria of the positive-feedback-driven dream state. The experiences may be enjoyable, but they are also completely out of touch with the world, with actual events. A normal life consists of a healthy balance between the two modes.

There is something to be said, however, for the quiet, reflective mode, working with the private thoughts created internally, without outside interference. This is where creativity lies. The solitary mind is often the inventive, ingenious mind. If we are never alone, we might never create. If we are never left alone, we might lose our ability to think unaided. To dream. To fantasize. And thereby to create and invent.

Perhaps sometimes we should do without specialized devices like the Teddy. Sometimes we should leave things as nature made them. Millions of years of evolution have led us to a very special niche. The human mind is very specially adapted to its environment. It is dangerous to tinker with it, whether through drugs, physical manipulation, or mind-control. It is too complex to be understood, and the effects of the tinkering are more apt to lead to harm than good.

A Teddy could be a wondrous thing. Maybe. But will solitary thinking disappear, along, perhaps, with great creativity and invention?

Moravec's Robotic Vision

It is easy to make a case for the synergy of human and machine, each left to do what it does best, each complementing the skills of the other. The problem comes when the machine takes over from the human, taking away initiative, forcing the human to serve as slave to the ends of the machine.

The extreme case of takeover is that advocated by Hans Moravec. In his book *Mind Children: The Future of Robot and Human Intelligence*, Moravec predicts the day when machine intelligence will equal or surpass that of people. Moreover, Moravec points out, a machine intelligence should never wear out, for it is based upon information, not mechanics or biology. Machines do break of course, but when they do, they can simply transfer their knowledge to new machines. The knowledge can thereby stay around forever, for knowledge leads an existence quite removed from the machines in which it resides.

That would be alright if Moravec were willing to stick to machines. But Moravec is after immortality: He wants human knowledge to exist forever. Not just knowledge, but minds. Although your Teddy may have started out as trusted aide and confidant, over time it could *become* you, at least if Moravec's vision were to come true.

Think of it this way. Project yourself into the future to a time when mechanical parts have reached a skill undreamed of today. If you damage a leg or arm, it can be replaced with a mechanical one, just as effective and functional as the original. In fact, each biological part of the body can be replaced by a mechanical one. Arthritis, rheumatism, broken limbs, failing hearing or eyesight are all things of the past, for as each biological structure fails, it can be replaced with its mechanical equal. Imagine growing old and replacing body parts as they fail, which might result in having a completely mechanical body. Hard to imagine, maybe, but possible.

But what about the brain, what happens there? Why we can replace that, too. Now this is tricky. If we simply remove the brain and pop in a new one, whether biological or artificial, what happens to you, your mind, your experience, your very existence? Gone. Your feeling of self is woven into the structure of the brain. You can't just pop in a new brain and continue to function. The new brain wouldn't have an identity, or at least, it wouldn't have your identity.

Moravec suggests we overcome this problem through a steady replacement of cells: Simply replace each brain cell with a mechanical equivalent, carefully adjusting it so the knowledge and structural information in the biological cell is exactly duplicated by the artificial one. You would never know the difference.

See how it works? I connect a computer up beside you in the operating room. Then I peer into your skull and replace one cell with a computer circuit. I carefully find all the connections of the cell and replace them with wires to the computer, and then I adjust the computer program so that it exactly duplicates the functions of that cell. You can do the judging because, fortunately, brain cells cannot feel pain. It is possible (and common) to do surgery on the brain while the patient is alert, conscious, and talking with the neurosurgeons. So, I replace your cell with the computer and then switch between the two—first to the real cell, then to the artificial one. "Notice a difference? Yes? OK, just a moment." I tinker with the program some more, then switch back and forth again. We simply keep doing this until you can't distinguish between the real cell and the artificial one. Then, zap the real one—it isn't needed anymore—and do the same for the next. Eventually you will be in the computer, not in the brain, and you can't tell the difference. The procedure guarantees that.

Of course, at some point, it would not be clear who "you" is. Which is exactly Moravec's point. If you can't tell the difference, why would you care? Your mind and all your experiences can live forever, moving to newer computers each time the old one wears out. Except that, well, Moravec doesn't really think much of emotions, so those got left out. And actually, he doesn't think much of the human mind either, so as long as he has it in information form and inside a computer, which we all know can do arithmetic and remember properly without error. Why not take advantage of these capabilities and use them to improve upon the mind? To make it remember better, do arithmetic better, think more logically.

Moravec's brain transplant scheme is very clever. Teddy takes over. Fortunately, the plan has many drawbacks.

1. It assumes that a single cell is independent enough that we could replace it, then go on to the next one. But its operation may be tightly linked with that of thousands of others (it is not unusual for one brain cell to make 10,000 connections to other cells), and the fact that it seemed replaceable

under rather special circumstances—the quiet of the operating room—does not mean that it will work correctly in other situations.

2. Single cells seldom provide essential information. Remove a single cell and there will be no effect on the normal operation of the brain. That, in fact, is how the brain maintains its reliability and ruggedness: When cells die, it seldom matters. It is cell assemblies that matter, and these may consist of hundreds of thousands or millions of cells. Replacement is no longer such a doable task.

3. The scheme ignores the chemical operation of cells. Basically, Moravec assumes that a cell is a cell, working in isolation, operating solely on information electrically conducted through its fibers and junctions. He completely neglects the chemical operations. The proper functioning of a cell depends upon the makeup of the chemical fluids in which it bathes. Not only that, but there is a vast chemical information processing system, huge ducts in the brain that convey important information through the composition of its fluids. Each duct bathes millions of cells.

When you look at a photograph or drawing of the brain, you will see large dark areas, empty space. These are the ventricles, the ducts that deliver fluids to large areas of the brain. The subtle, complex mixtures of chemicals that result have major impact on the operation of the brain cells.

Get agitated and the hormonal glands squirt their output into the ducts, which rapidly carries them to large areas of the brain. The chemical signals can be remarkably selective, affecting only cells that have specific receptors for the particular chemical. But those cells then change their operation in a manner that depends upon the nature of the chemicals and the circumstances, in ways that are not yet understood.

Moravec's notion of replacing brain cells one by one will miss this most important part of brain functioning. In fact, because the new brain that Moravec is creating does not have these chemical pathways, it will never work the same as the old. On top of that, even if Moravec suddenly begins to understand the chemical structure of the brain, he will have difficulty reproducing it. His scheme works because it assumes that each element works independently of the others, so that if you replace it with a different device that has the same characteristics and the same connections, the system will not notice any change. But that isn't how the system works at all: The cells do not work independently of one another, especially when chemical communication is taken into account.

4. Finally, replacing the brain cell by cell would simply take too long. That is like saying that it is possible to win at chess simply by thinking through the implications of each move. True, but there are so many implications that even the world's fastest computers could never manage it in many lifetimes. That has been known for decades.

There are at least 10^{11} neural cells in the brain, or 100 billion (some estimates are 10^{12}, which is ten times more). If each cell has to be replaced by asking the person if any difference can be noticed, it will take some time to allow for a question, a reflection, and an answer. Let's assume it can be done at the average rate of one second per cell. One hundred billion cells means one hundred billion seconds. That's $1\frac{2}{3}$ billion minutes, almost twenty-eight million hours, over a million days, or more than three thousand years. All in the operating room—after all, you wouldn't want to stop with half your brain in your head, half in a computer.

So the scheme won't work. Thank goodness.

The Future

What I want is all of the virtues of machines and none of the disadvantages—the scientific version of eating my cake and having it too. After all, if I carried my own information bank, my own Teddy, with me at all times, with me in control of the on-off switch, what are the deficits?

Alas, technology is always a double-edged sword, always show-ing two faces to the world. Every benefit has its accompanying draw-back. One possible deficit is simply the danger of the tuned-out world. Look around you today and you can see the early beginnings. All those people, walking about with earphones strapped to their ears, wandering through the world. They are tuned in to their own senso-rium, tuned out of ordinary human discourse or interaction with the environment.

I have mixed feelings about this Teddy. I can imagine the good things. I can fear the bad. But in actuality, I have little choice in the matter. It is coming into being. Yes, some form of Teddy will be with us, but the exact shape or form cannot yet be predicted. But whatever form it takes, I predict it will be both a boon and a loss. It will lead to reasonable and sensible fears and hopes. And if we humans are not sufficiently intelligent about its design, functionality, and use, it will forever alter our lives in ways we do not want.

7

How Long Is Noon?

AM AND PM. One means morning and the other afternoon, right? Everyone knows this, and everyone knows which is which—that is, everyone who has spent the several hours of instruction required to have learned the difference, and the tens of hours required to learn how to tell time. The device we call a "clock" is really rather peculiar, an artifact of a very old-fashioned technology. Three hands revolving around a common point, with two different numerical scales, one graduated from one to twelve (even though the day has twenty-four hours, not twelve), the other two hands graduated from one to sixty. No wonder this system causes children to have great problems and adults occasionally to err.

And if AM is morning and PM afternoon, what are 12:00 noon and 12:00 midnight? If midnight is in the evening, it ought to be PM, or does it mark the first instant of the new day, in which case it is "before noon," so it should be AM?

And if noon is the division between morning and afternoon, the place where we start counting afternoon time over again, shouldn't the hour after noon be counted as zero, not twelve? Shouldn't ten minutes after noon be 0:10 PM, not 12:10 PM? The same goes for midnight: We start our time over again at midnight, so that first hour of the morning is the zero hour, not the twelfth hour. Calling it twelve makes it seem like part of the preceding day. Ten minutes after midnight is ten minutes into the brand new day, so shouldn't it be 0:10 AM? Why is it still counted by yesterday's numbers, as 12:30 AM? Why do the numbers start over again at 1 AM or 1 PM, instead of at noon and midnight, for aren't the two "twelves" the magic hours of time telling and the important dividing marks?

Which is earlier, 11:30 PM or 12:30 PM? The answer is 12:30 PM, which is eleven hours earlier than 11:30 PM. Why? Consider the following times:

6:10 AM
8:15 AM
12:40 AM
1:50 PM

I have carefully ordered the times. Nothing special, right? Well, there is something special: The order is wrong. I should not have put 12:40 AM after 8:15 AM. The correct ordering is:

12:40 AM
6:10 AM
8:15 AM
1:50 PM

If this seems wrong and counterintuitive, blame it on our archaic time-telling system. The best way to handle the problem would be to have time start all over at midnight, so that times after midnight are zero-hour times. Not 12:30 AM but just plain :30, or perhaps 0:30. Then the times would read:

0:40 AM
6:10 AM
8:15 AM
1:50 PM

Both numerical and time-telling order would thus coincide.

In fact, let's consider the difference between twenty-four-hour time (as used throughout Europe and by the U.S. military) and the twelve-hour AM/PM style we use in the United States.

AM/PM Time	24-Hour Time
Time-of-Day Order	
12:40 AM	0:40
6:10 AM	6:10
8:15 AM	8:15
12:30 PM	12:30
1:50 PM	13:50
6:10 PM	18:10
Numerical Order	
6:10 AM	6:10
8:15 AM	8:15
12:40 AM	0:40
1:50 PM	13:50
6:10 PM	18:10
12:30 PM	12:30

In the left-hand column I have put times in the twelve-hour, AM/PM format, while in the right-hand column the same times are given in twenty-four-hour format. In the top part of the table, the times are listed by time order and in the bottom part by numerical order, in both cases according to the times for AM/PM format. To my eyes at least, the AM/PM format looks strange, for the time order differs from the numerical order. To determine that the time order is correct (or the numerical order wrong), I have to think about the meaning of the times for 12:40 AM and 12:30 PM. The twenty-four-hour time makes the most sense, for here, time order and numerical order coincide. It is easy to tell that the bottom right-hand column of times is wrong, for it is neither in numerical nor in time order.

How Shall Noon Be Labeled?

What about noon: How shall it be labeled, AM or PM? In a clever essay on the topic, the engineer Henry Petroski argues that 12 noon should be labeled 12 M, for after all, the history of AM and PM is that they mean *Ante Meridiem* ("before the middle of the day") and *Post Meridiem* ("after midday"). Noon *is* the meridiem, the midday, so it is neither before nor after. It should be labeled *M*: 12 M.

88

Yikes! That is putting principle first, and damn the consequences. I hate to disagree so strongly with one of my favorite authors, but I splutter at the thought of using M to mark noon in order to distinguish it from midnight! In English, if any letter is to be used to denote noon, it should be N. M ought to stand for midnight.

The mark of "M" for noon makes historical sense, but it makes practical sense only if the everyday user of time understands the original meanings of the terms AM and PM. But today, most users think that AM means morning and PM afternoon, and are puzzled over the origin. The terms AM and PM may have made sense in the age when only intellectuals cared about the exact time and all intellectuals spoke Latin, but not today when time is of critical importance to all in a technological society, when few know Latin (or even realize that AM and PM derive from Latin), and when noon is a frequently specified time. In the English-speaking world, 12 M would surely be thought to mean 12 midnight!

The specification of time is an evolutionary process, one of slow irreversible changes. Evolution slowly changes life, and although we still have vestigial reminders of the past, their meaning has long since disappeared: The fact that the M of AM means midday or meridiem is lost to most people as surely as the function of the appendix is lost to most bodies.

In countries that speak Romance languages, M might still mean *noon*, just as A will still signify *ante* or *anti* and P *post*, but what about all those countries with different language bases? Should each language use a different symbol for noon? And what about midnight? If noon is M, what is midnight?

The labeling of noon and midnight are indeed problems. What do I suggest? Two things. First, I myself always use the full word "noon." Twelve noon, I say, or 12 midnight. No abbreviations: they are dangerous. The historically correct abbreviation of "M" for noon is guaranteed to give trouble however it is used: Some will think it means midnight, others will think meridiem.

So, I would label 12 noon as 12 noon. I would also combine this with starting the day's time count at midnight, with the hour between midnight and 1 AM called the 0th hour. We thus can finally get the AM/PM time in a format whereby numerical time coincides with clock time.

Modified AM/PM Time	24-Hour Time
Time-of-Day Order	
0:40 AM	0:40
6:10 AM	6:10
8:15 AM	8:15
12:30 Noon	12:30
1:50 PM	13:50
6:10 PM	18:10

Alas, Petroski disagrees, for the use of the English term "noon" violates the tradition of using the Latin symbols AM and PM. Thus he complains that "many an itinerary, program, or schedule will employ a curious mix of Latin and English by listing morning hours as 'A.M.,' noon as simply 'noon,' and afternoon hours as 'P.M.' "

Ah, the technologist versus the humanist, except in this case, we cross roles. He, the engineer, is the traditionalist, wanting to maintain old traditions and distinctions, even in a world where they no longer make any sense. I, the social scientist, want to take the world as it is today and not use ancient appendices that no longer have functionality. In fact, you can notice another sign of the difference between us: What he calls "A.M." and "P.M." I call "AM" and "PM." The difference is in those periods. To Petroski, the humanist, "A," "P," and "M" are abbreviations and therefore must be followed by a period. To me, the terms "AM" and "PM" are technical symbols, the one meaning morning, the other afternoon, and as complete, whole symbols in their own right, I see no need to use periods between the components. Just as everyday folks have lost the Latin etymology, and just as the abbreviation of "M" would be counterproductive in a society where "M" historically stands for noon but today would mean midnight, we no longer need to keep the periods. Nonfunctional design disappears. Nonfunctional evolution need not disappear.

The same is true for many of the abbreviations of daily life. AM and PM: What do the letters stand for? You probably know what BC stands for in calendar years, but do you know what AD denotes? How about AC and DC in electricity? Or AM and FM in radio? Does knowing any of these abbreviations help with understanding the AM and PM of time? Answer—no. In fact, so little do these other abbreviations convey their separate words (even those from English, not Latin) that I hear engineers talk about AC voltage and current, unaware that this is either nonsensical or redundant. (The first means Alternating Current voltage—which is nonsensical; the second means Alternating Current current—which is redundant.) Then we have AM modulated radio signals (which means Amplitude Modulated modulated).

The point is that abbreviations can become functional words in their own right, with the meaning of the separate components lost, except to the student of language. To insist upon maintaining the historical heritage even in the face of resulting confusion and possible error is to fail to recognize the need for everyday design to accommodate real human behavior.

How Long Is Noon?

While I am at it, let me address yet another critical question for those of us who are clock-watchers. Having decided that twelve of midday is neither AM nor PM but noon, how long does noon remain noon? Perhaps it should be one second—only the second following 12:00 would be labeled "noon." But why one second? That is an arbitrary period of time. Why not let the entire hour be the "noon hour"? Or maybe noon is an arbitrary instant, lasting for such a short duration that the transition from before to after can't be experienced.

The suggestion that noon has some duration is clearly one rooted in technology. If a clock kept its three hands stationary for a second, it would hold the indication of noon for an entire second before moving the second hand away from the 12. On such a timepiece it would make sense to display "noon" for that entire second.

But if we are to let technology determine the duration, then the length of noon will vary with the level of technology. Modern mechanical watches move the second hand five or more times a second, so for these watches, noon would only last one-fifth of a second. On a digital clock, time is divided into discrete times that reflect the oscillator frequency of the underlying electronic circuits.

But whatever the choice, noon will last only a few microseconds, at best.

How long does noon really last? In fact, noon exists for ε seconds, where ε, the Greek letter "epsilon," is the standard mathematical notation for a small quantity. How small? As small as possible, for the convention is to let ε shrink in size to get as close to zero as can be imagined without actually becoming zero. To the practical engineer (such as me), noon therefore exists only for an epsilon of time, which means that for practical purposes, it doesn't exist. A time that is specified as ε is instantaneous. And so the digital clock is correct in switching immediately from AM to PM, with not even the slightest pause to recognize the existence of noon.

How long is noon? How long is midnight? They don't exist as durations, they exist as landmarks, places to anchor the day and night, a time to meet someone or to do something, but not as durations. You can meet someone at noon, but you cannot accomplish anything at noon, for noon doesn't last very long at all.

8

Real Time

THE phrase "real time" is a frequent one among scientists and technologists who deal with computation. Real time? As opposed to what? Unreal time? What on earth is real time? Or better yet, what is unreal time?

Real time is defined as the time during which events actually occur. If you drop a plate on the floor and it smashes to bits, it does so in real time. Now, isn't that illuminating? Aren't you glad that technology has given you this great insight into the nature of the world?

The term does make some sense when we compare how long it takes to compute something with how long the event itself takes. The need for the term "real time" came about when we discovered that our fancy computing machines couldn't keep up with reality. Computer programs that predict the weather might take twenty-four hours to make a prediction one hour ahead of the time they started. The answer comes twenty-three hours too late: The computer takes so long that we say it does not work in real time. It works in its own time frame.

Most computer programs that try to simulate a complete piece of the world cannot keep up. Programs that try to understand spoken sentences lag far behind the actual speech. Programs that try to mimic in exact detail the movements of animals fail to keep up with reality. And we ourselves can fail, as when we listen to a speaker rapidly reading some complex text: We cannot follow the implications as rapidly as the speaker can state them. Mind you, the speaker cannot think of them so quickly either, which is why the speech has to be carefully prepared ahead of time and either written or carefully practiced and learned. Normal speaking and listening are rather well matched to the capabilities of both speaker and listener.

We can speed up or slow down time through art and technology. A play might take an hour to present five minutes in someone's life, or it might pass over generations in a few minutes. Similarly, a movie or video camera can take one photograph per hour of a growing plant or other slow-moving scene, which we can then play back at twenty to thirty times the actual rate. Or we might photograph a rapidly moving object several hundred times a second and then play it back at one-fifth or one-tenth the actual speed, allowing us to see the details. Slow motion—slowed-up time. Books change the passage of time, for the writer, the reader, and the events themselves are totally disconnected. The writer may take days or weeks to write a single page, while the reader may pass over it in a minute, or spend hours reading and re-reading it, or skip it altogether. The event being described may be a static, unchanging one, or a rapidly changing one, or even an imaginary one: It makes little difference.

Music is usually listened to at the same rate it is performed, because the very nature of sound is such that if slowed down or speeded up, its experiential qualities differ. But that does not stop us from a slow perusal of the score, or from mentally replaying the sounds at whatever rate we choose.

But in all these instances, what exactly do we mean by real time? The time of the event? Sure, that is a real, true time. The time of the experience? Sure, that is also a real time. And if the experience takes a different amount of time than the event, which is more real?

Human Abilities and Real Time

Real time may make sense for the arts or technology, but does it make sense when talking about human performance? Recently, I was at a meeting discussing the difficulties of getting computers to perceive visual scenes—a field called computer vision. One of the participants remarked on the powerful, sophisticated properties of human vision that could see images in real time. Hmph. Yes, the human sensory systems are all marvelous, but what is this real time stuff? Things we can see are by definition "real time" and things we can't see aren't.

If we wanted to play this game, we could comment on how slow is human vision. Moving objects appear blurred and we can't see details on them. We can't read the lettering on a moving ball, and we can't even see a speeding bullet. Does this mean that we can't see in real time?

Did you ever trip while walking? Why? Because you didn't compute fast enough—your brain wasn't following in real time. If I suddenly turn and throw something at you, I will probably hit you. Obviously, if you could compute in "real time," you would dodge and I would miss. The notion is that the computations needed to move the hand and body have to be done in "real time." If you take longer than "real time," the ball will get you. How wonderful the human who can compute in real time! How pitiful that we cannot do everything in real time. Bah!

By definition, real time is what humans do. So it isn't wonderful that we can do it! Sure, we can catch a ball in real time, but that is why we play ball. We wouldn't play ball if we couldn't do it.

When talking about human capabilities, real time is a silly concept because we have no choice in the matter; events take as long as they take. Moreover, the nervous systems of all animals have evolved to ensure each animal's survival, which also means the system allows the animal to respond in appropriate time. Do animals respond in real time? Of course, by definition. That is, they respond to events in the amount of time they take to respond to them, which is by definition real time.

If physical events were to happen much more rapidly or slowly than they do, then presumably our nervous systems would have evolved differently. If one animal has much faster responses than its competitors, then it will triumph—unless its competitors evolve into faster-responding animals themselves. All in all, evolution works so as to keep the time factors roughly equivalent. We respond in whatever time we need to meet the demands of our environment.

How fast does the nervous system have to work? Fast enough to deal with the most common movements. Perhaps fast enough to see a falling body coming from, say, treetop height. Perceptions of speed are, to a large extent, dictated by the events around us. The amount of time it takes something to hit the ground from a fixed height is inversely proportional to the square root of the force of gravity. As a result, the force of gravity is probably a major determinant in how fast our nervous system has to work.

The moon's gravitational force is 1/6 that of earth's, so it takes an object 2.4 times as long to fall ($\sqrt{6} = 2.4$). Jupiter's gravitational force is 2.6 times that of earth's, so an object falls in 60 percent of the time it would take on earth. Does this mean that if intelligent life had evolved on the moon, thought would take place more slowly than on earth? Or that intelligent thought on Jupiter would be much faster? Could be.

How is it that the nervous system works at just the right speed, pacing the natural events of the world so that the most essential operations are indeed done in real time? The answer can be found in evolution, which provides a mechanism for slow, steady adjustment, matching animals to the environment and the other organisms with which they compete. Natural evolution is slow, however, so it is easier to see this process at work in an artificial situation: the artificial evolution of games.

Games evolve through deliberate, purposeful change. Games require change to maintain an effective match with their environment, with the people who compete in them, and with those who watch. Games must be challenging for even the most expert player, yet remain playable and interesting for beginners. The nature of the game is important for both player and spectator, so it should move quickly enough to be of interest to those watching, but not so quickly that it cannot be followed or performed.

Baseball tends to be a mysterious game to those who did not grow up with it. But for our purposes today, we simply need note that it is primarily a game between the pitcher who throws a 2.85-inch-diameter ball at a batter, who has to hit it with a stick whose thickest end is 2.75 inches. Today, the pitcher stands on a small mound of dirt 60.5 feet from the batter and throws the ball between 70 and 100 miles per hour. The ball is constructed according to carefully stated rules that specify its composition, weight, "bounceability," and even the construction of the seams on the ball. (Because the pitcher spins the ball at around 2,000 revolutions per minute, the spin and the seams make the ball's path curve, drop, or "float.")

In baseball's early days—around 1840—the pitcher stood only 45 feet from the batter and threw the ball underhanded. The batter was out if he hit a fair ball that was caught in the air or on the first bounce. The batter was also out if hit by a ball thrown at him when he was off base.

Over the years the rules were changed in order to equate the skills of batter and pitcher. Some changes were made to make the game more interesting for spectators. Some adjustments made it harder for the batter, some for the pitcher. The spitball was considered "unfair," because the ball seemed out of control. But then again, the first bunts were thought to be unfair. The strike zone was made larger (to ease the task on the pitcher) and then smaller again. The pitcher's mound kept getting lower and lower, making the pitcher's job harder. The sacrifice-fly rule came and went, and came again, as the continual fine-tuning over the speed and mechanics of the game continued.

All games evolve in a similar manner, all games that survive, that is. The task is to keep that deliberate balance between opposing sides, in part, you might say, constrained by the limits of "real time."

Technological Time

Today we think of time as measured by the clock: hours, minutes, seconds. But this is physical time, an invention of technology, not a natural concept. In fact, there are a number of different kinds of time: real, psychological, physical, and political. Real time is the duration of a physical event. Psychological time is the subjective experience of its duration, how long it appears to take. Real and psychological times can be very different.

Physical time is an arbitrary quantity, defined over the years by physicists and based upon counting some recurring physical event. In the beginning, time was defined physically by the passage of the moon, stars, and sun across the sky. The marking of time eventually moved to counts of simpler, manufactured objects measuring drops of water or grains of sand, or the number of swings of a pendulum. Today physical time is measured by counting the cycles of vibrating atoms.

Finally, political time is defined by convenience, a set of international standards developed so that everyone can agree upon how long something takes, or for that matter, upon what time it is at any particular instant. Psychological time does not work in such orderly, evenly spaced ways. Time passes at a different rate when we are asleep from when we are awake. Dull activities pass more slowly than do exciting ones, although in our memories, exciting events may have seemed to have lasted much longer than the dull ones. Thus some people who have been in automobile accidents comment on how long those last few instants before impact seemed: Psychological time was stretched.

The world of technology, however, moves according to things that are measurable, and subjective, psychological time is not readily measurable. Worse, two different people might experience two very different times for the same experience. As a result, an artificially contrived definition of time has come to dominate, supporting the notion of time as continuous, regular, and unvarying.

Some readers are sure to cringe at the statement that our definition of time is artificial. Time is a physical property, they will say. It is no more arbitrary than space and distance. Exactly, I respond, for the way we measure space and distance—in three orthogonal axes—is also arbitrary.

The real point, however, is not to argue the merits or demerits of physical quantities. It is simply that psychological, perceived, human time is quite different from the mechanical technologically defined passing of years, days, and nanoseconds. We have come to redefine ourselves through mechanical time, even though that is not the same as psychological time.

Time-Telling

In the early days of time-keeping, we marked the day by the progression of the moon, sun, and stars across the sky. Today we still use landmarks to help define the time: sunrise and sunset are two, the high-noon sun an obvious third. Sundials allow us to see the changing path of the sun. The simplest sundial is simply any vertical object that can serve as a shadow-caster. As the sun moves east to west across the sky in the northern hemisphere, the shadow-caster produces a shadow that moves in a curved path from west to east, curving around the north side of the shadow-caster. In the southern hemisphere, the shadow curves around the southern side of the shadow-caster. The shadow is longest at sunrise and sunset, shortest at midday. The shadow's direction of travel in the northern hemisphere has come to be called "clock-wise." The direction we call clockwise is therefore just as arbitrary as the length of time we call "an hour": If sundials had originated in the southern hemisphere, clockwise would mean rotation in the opposite direction.

There is less daylight in the winter than in the summer. In my hometown (San Diego, at 33° latitude), on the shortest day of the year, there are almost ten hours from sunrise to sunset, whereas on the longest day, there are around fourteen. San Diego is in the southern part of the United States, the same latitude as northern Africa. Locations closer to the poles have even greater differences between the amounts of daylight hours in summer and winter.

If the day were divided into an equal number of hours from sunrise to sunset, in San Diego the summer hour would last almost $1\frac{1}{2}$ times as long as the winter one. This was the way it was when time depended upon sundials. Even after the first mechanical clocks were built, there was an attempt to mimic this aspect of nature, so that summer hours on mechanical clocks actually took more physical time than winter hours. This complicated the mechanisms enormously and, as you can easily see by looking at how we keep time now, did not last. To me this seems a shame. In the heat of summer, wouldn't it be nice to have more physical time in which to do an hour's worth of work? In the cold of winter, it seems more appropriate to hurry to do the same work.

If we were to divide the span of light from sunrise to sunset into a fixed number of segments, there would be other interesting properties. It would take less physical time to drive between two cities or to fly across the country in summer than in winter. The poor student stuck in summer school might complain with some justification that the hours seem to crawl by at a too leisurely pace.

On the other hand, predictions of how long it would take to do a job might very well stay the same from season to season. The carpenter promises the job will take six hours, and six hours it does take, whether it be summer or winter. The summer job allows more time to rest, to ponder the flowers in the yard and the clouds in the sky. The winter job means continual work with nary a break for relaxation. Seems a civilized way to live.

Of course, living time by the sun and season of the year would work only as long as we did not need to synchronize activities with people who lived far away. The long summer hours of people living in the northern hemisphere would coincide with the short winter ones of people in the southern hemisphere. If I wanted to call someone in Australia, I would have to make sure we agreed upon whose clock we would be using. In fact, the length of an hour in southern Europe would differ from its length in northern Europe. Mexican time would not be the same as Canadian time. Sun time works only locally: Once we try to synchronize events across a geographical region, we must accede to mechanical time.

Psychological Time

Psychological time is measured by the things we experience. In fact, the memory of our experiences for an event long after it has passed may very well differ from the experiences at the actual time. Drive a car across the United States and there are stretches of endless, unchanging vistas that seem to take forever to pass. Or attend a graduation ceremony as a member of the nonparticipating audience and sit mindlessly listening to the speakers droning on and on. But a year later, the dull parts of the automobile trip or the graduation ceremony are hardly remembered. In our memories the significant events are what dominate the trip.

Subjective time depends very much upon the interpretation of the events. Note the discrepancy between experienced and remembered duration. Suppose I have you sit in a dull, eventless room for an hour: no books, no music, nothing to occupy your mind. During the event, the relative lack of notable events makes time progress painfully slowly. A year later, you may remember the experience, but the lack of remembered events means that you will probably remember it as taking but a brief time—even if you also remember that you were horribly bored.

Psychological time is a personal experience, more dependent upon how you interpret the events than anything else. Much has been said about the difference between Eastern and Western time, with the notion that Eastern cultures have a different experience than do Western ones. Perhaps the Tibetan monks have different notions of time. Indeed they do, but the difference is not between East and West, it is between subjective and physical time.

Technological societies tend to be dominated by the clock. This has been caused by the need to synchronize activities across many people, some of whom are separated in distance. The clock defines our events. We start and stop work or school at specified times. We eat at meal time, not when we are hungry. We wake up to the sound of an alarm, not when we are rested. Everything is done according to schedule. Wasting time is considered a sin. In other words, physical time is governed by the relentless progression of the clock, despite the fact that it conflicts with psychological time, despite the fact that it conflicts with human biology. The human body naturally follows a daily rhythm with two sleep periods: the major one at night, when most of us do indeed sleep, a minor one in early afternoon, when, well, many of us also do indeed sleep. (A sleep researcher once told me that she was careful not to schedule any of her talks for early afternoon, especially just after lunch, because she knew that much of the audience would be sleeping, or at the least, drowsy.) But technological society is ruled by time and money, wishing to waste neither.

There still are countries that practice the afternoon nap, where shops, banks, and offices close in mid-afternoon. I once thought this a peculiar custom, but no longer. Today I tend more and more to think of these countries as civilized, the rest of us as mechanized. Doesn't it make sense to attend to the body's natural rhythms and needs? Mind you, these countries still live by the clock, and the starting and stopping of the afternoon break follows an established schedule. But the scheduling allows an interesting respite, allowing those who feel the need for a rest break to have one without feeling like social outcasts.

Many of our most pleasant moments occur when we lose all track of time, captured by the flow of activity, engrossed by a gripping novel, or by a theatrical or musical performance. When this capturing occurs, the subjective experience is that the rest of the world disappears. There is no sense of time, just intense concentration and involvement with the activity. The capturing can be work or play, it doesn't matter. It can take place only if the environment permits it, and one of the most effective ways to prevent the state is to have interruptions. Or to look at a watch.

Subjective time can be experienced only if you attempt to experience it, if you ask yourself, "How long has this been going on?" If the event is truly captivating, then you don't even ask the question.

So, how long is time? As long as you would like it to be. How long is real time? As long as it takes.

9

Nature's Packaging

MODERN society produces an amazing variety of stuff, then delivers it to market, packaged so as to protect, inform, titillate. Consumers accept the offerings, but not always with pleasure. Some packages promise more than they deliver, such as the large colorful package that opens to reveal empty space, marginal contents. Others seem more designed to protect than to be used—the metal sardine can that resists all attempts to be opened, the plastic bag of nuts that requires teeth, knife, and savage attack. Some packages attack the opener, such as the beer can that squirts the user in the eye. Some turn into deadly weapons after being opened, such as the jagged edges around the soup can. And some of this packaging resists destruction, as proved by the various plastics that survive the garbage truck, the city dump, and the incinerator to litter the environment and wash up on our beaches.

Nature too must package its perishable goods. The variety of methods used rivals that of people and in many cases is surprisingly similar. For in this situation as in many others the demands of the task dictate the solution, and the task requirements of nature and of the modern industrialized human are surprisingly similar. Nature and the modern retailer have similar goals. They must advertise their wares, attract consumers, and protect the wares during transit. Moreover, the product has to deliver as advertised to its consumer if it is to keep its base of customers.

As every schoolchild knows, many plants use animals to aid in reproduction. Colorful flowers attract birds, butterflies, and bees that aid in the pollination. Attractive fruits with sweet, fleshy exteriors surrounding hard, indigestible pits are designed to be attractive to animals who will eat the fruit and then carry the pits around inside them until they finally evacuate the residue at some distant location. The flesh of the fruit is designed to be tasty to the animal, the pit is designed to resist the digestive fluids and emerge from the animal unscathed by its strange form of transportation. The fleshy part of the fruit is therefore simply an advertising gimmick, a tidbit that entices the animal to do the real task of transporting the seed.

The coconut has to survive a long fall from the top of the towering palm tree, then be transported for long distances by river or ocean until it finally washes up at some distant location and propagates itself. Some fruits—the banana, for example—are designed to be easy to open, others to resist opening. Some must be consumed immediately, some are designed to be stored, perhaps over the winter. In many cases, nature's packaging also serves the needs of the modern storekeeper, the natural packaging of many plants allowing them to be shipped long distances and displayed in stores with little or no extra effort, aside from cleaning (the better to display nature's natural coloring).

But nature does not always share our needs, or our views on consumer protection and product safety. Some of nature's products are dangerous to handle (harmful plants such as poison ivy and oak abound). Some poisonous plants are difficult to distinguish from others that we cherish as food—consider the mushroom and toadstool for example, where poisonous species are barely distinguishable from tasty treats. Some packages hide the quality of the stuff inside, as with the coconut or pineapple, or even the orange. Some are easy to open, some difficult. Some have a residue—the peel—that is difficult to dispose of. Some plants have delectable exteriors and poisonous interiors. And being able to tell how juicy or ripe or firm a fruit or vegetable is from the outside requires the special art of the expert. Favorite foods such as coffee, tea, and cocoa are in fact drugs and if made artificially, would be subject to all sorts of regulations and prohibitions. Some natural stuffs, such as some forms of poppy seeds and marijuana, are in fact prohibited.

Nature works its ways very differently than do people. It has no goals, no morality. Its time frame is also very different. In some sense, everything is a temporary phenomenon that will be corrected eventually. The earth sustains major events: huge volcanic eruptions, sizable earthquakes, the formation and movements of continents, and bombardment by huge meteors. Each event makes its mark upon the earth, but to nature, time is measured in millions of years and so the marks eventually pass.

The disposal of waste is not always part of nature's plan. Coral is an animal, not a plant, but even so, it is a monument to the indestructibility of nature, for the mounds of dead bodies and skeleton remains of the past exceed the size of even the largest of today's garbage dumps—the discarded bodies form piles large enough to rise up from the sea and form islands. Animal bones litter the ground for millions of years. Nature has even turned some trees into stone, the better to preserve the litter, with petrified forests scattered next to huge erosion pits, perhaps the most famous being located in the deserts of the southwestern United States.

Today we would not allow the conditions to exist that led to the Grand Canyon. Any reasonable environmental impact report would have identified the Colorado River as a scavenger that would eat away the rocks, especially the soft sandstone. "Why, if we permit this stream to continue in this way," I can hear the federal examiner saying, "this entire region of the country will be one vast wasteland. The river will cut into the land, ten, maybe one hundred feet. It will be impossible to build roads. And think of the silt and destruction of native environment and the change in habitat for the existing animals. No, this river cannot be permitted to flow in this fashion."

But such sensible analyses have no impact on nature. We cannot prohibit coral from occupying the ocean unless they change their bodies to be decomposable. As a result, coral has destroyed vast sections of pure, uncluttered ocean, creating land masses where there was nothing but empty ocean, thus violating the homesites of natural fish and ocean vegetation. As for the Colorado River, the destruction has exceeded even the worst predictions of the environmentalists: erosion a mile deep extending over hundreds of miles. It should be a national disgrace.

But we do not prohibit nature from the same activities that we prohibit in ourselves. Instead, we do the reverse. We take pride in natural litter, inhabiting the coral dumps, making pilgrimages to petrified forests and canyons, but it is litter and waste nonetheless. Perhaps

100,000 years from now tourists will flock to marvel at our open pit mines, our mounds of automobile tires, and the debris from junkyards. With time, even discards acquire natural beauty.

Plants and animals evolved jointly, each forming a dependent relationship with the other. The hummingbird developed its skills at hovering and its long, thin beak in synchrony with the development of the flowers that required such attributes. The taste senses of fruit-eating animals evolved to be sensitive to the sugars of fruits. After all, the sugar content of a fruit would not be attractive to an animal unless it enjoyed and was sensitive to the taste. Similarly, the colorful display of flowers and birds would serve no purpose if the relevant animals were color-blind. Most animals are, in fact, color-blind, but not fish, birds, and primates. Nature's packaging was not developed as an isolated activity. No, it was developed jointly through the development of the animal species that it serves.

A similar story can be told of manufactured goods: The goods and the needs of consumers evolved jointly, through mutual accommodation. Today most homes have hot and cold running water, electricity, and heating. Years ago, these were foreign concepts. Electrical products would not work without households adapted to power them. Hidden behind the walls of the modern household are a mass of wires and pipes conveying electrical power, telephone, and television signals, available to the consumer at standardized connectors and outlets. We take these for granted, but were it not for national standards for the television and telephone connectors and the electrical plugs and sockets (along with standards for the signals and the voltage and frequency of the electrical power), we might have difficulty knowing whether a newly purchased appliance would actually work within the home. If manufactured goods are not designed to fit the facilities available to consumers, they are useless. If a product is too specialized, it may never receive sufficient customers to make it worthwhile.

Nature often takes just the opposite tack: It goes for the small, specialized product, with limited distribution. Some plants have fruits that are available only to a limited number of animals. Some parasitic life can survive only with the aid of a single species. Moreover, nature is willing to take thousands of years to accomplish this specialization. Small specialized niches of species are nature's way of doing things.

Human manufacturing can afford neither the time frame nor the specialization of nature. The hummingbird, the giraffe, and the anteater are all exquisitely tailored for the way their food is packaged. We humans can't afford to be so specialized. Here is where we differ from nature. Nature needs to protect its individual species by having specialized products. Suppose nature were suddenly to establish an international standard for fruits:

> *Official Notice:* Effective January 1, all fruits will be packaged in small round spheres of designated size, located a fixed distance from the ground. Poisonous foods will be given distinctive packages.

For nature, this simply wouldn't work, in part because species exist in competition with one another, and deception and hazards are nature's way of providing for survival.

Some manufacturing companies seem to side with nature's manipulative ways of survival. To the company, life is a deadly, competitive battleground, one where the losers do perish. A company would prefer to inhabit a specialized niche with no competition, but only if there were sufficient market for its products. To the company inhabiting an environment with fierce competition, deceptive, or at least "cleverly labeled," packaging may seem like the difference between survival and failure. It is clear that in the battle for survival and comfort, a company's perceived best interests are not always the same as those of its customers.

10

Evolution Versus Design

PHYSICAL artifacts are designed. People evolve. There is a vast difference between the two. Or is there?

The standard story goes like this: Design is optimization, the result of careful analysis of the problem and a solution making use of the sophisticated tools of design, engineering, manufacturing technology, numerical calculation, and computer simulation. Evolution, on the other hand, proceeds through slow, steady steps, each just a slight modification of what has come before. Sometimes the steps occur in a rapid burst, so that new species with new behaviors develop quickly by evolutionary standards. But the changes take place blindly: There is no way to know beforehand whether a modification will aid or diminish the chance of survival.

The difference between evolution and design seems large. Evolution is unguided, each new species advancing slowly by human time scales, and then only at the expense of others. Species are competitive, and the one most suited for the environmental niche tends to dominate at the expense of others less suited. Design, on the other hand, is guided through conscious, carefully crafted attempts to improve upon what has gone before. New leaps are not only possible but encouraged. If a previous design proves to have shortcomings or flaws, the designer can eliminate them completely from the next generation. Not so with evolution: There is no way to do away with past mistakes and start over, except by very slow incremental changes that might, eventually, eliminate the last trace of the flawed attempt. Evolution carries history with it, and each new adaptation must build upon the historical record.

But are evolution and design so different? Let's see.

Is Design Different from Evolution?

Evolution is not like design, but is design that different from evolution? Does the designer really have complete freedom to build anew, to look at the problem and then, starting off with no preconceptions, to solve the entire design problem? No, of course not. Design is much more like evolution than most designers would like to admit. In fact, one scholar has tried to equate the whole process of technological change and invention with the evolutionary process.

Why is this? Why isn't design free to follow the imagination and talent of designers, to go off suddenly in brilliant new directions, or to eradicate the failures of the past? Because designers are people, and people's imaginations and knowledge are limited. Breakthroughs and new approaches are rare. In fact, there is a good, substantive reason design is really like evolution: Knowledge is evolutionary.

The most essential tool of the designer is knowledge: knowledge of goals, intentions, and desires; knowledge of possibilities and methods; knowledge of technology and science; knowledge of tools and methods. And knowledge is cumulative. Our individual knowledge builds upon what we have learned previously. Society's knowledge is built from the knowledge of its members and the history of its institutions. Cultural knowledge proceeds in slow increments over prior states. Knowledge can expand upon, modify, and even contradict what has gone before, but it cannot erase its history and record. And major new advances in knowledge are difficult to come by, difficult to assimilate, and difficult to apply.

There are differences, to be sure. Knowledge can be directed. It can deliberately be steered in one direction or another. New ideas can be advanced much more rapidly than can new evolutionary directions. And old, discredited knowledge can be repudiated and fall into disuse faster than can the biological equivalent. But even so, it is surprising how slow and difficult is the process of design, how unlike the stereotype of creative insight, how much more like the slow, trial-and-error process of evolution.

The applied side of my research is concerned with the human side of technology, the so-called human interface. The phrase "human interface" refers to the part of the technological system that interacts with the person—the knobs, lights, meters, gears, computer displays, buttons, and pointing devices that form the "interface" between machine and human. In this area of applied research, my colleagues are fond of bemoaning the slow rate of progress, casting jealous eyes upon other design areas. Interface design is especially difficult, goes the complaint, because it involves the human, and science doesn't know enough about human characteristics to make design involving human capabilities into a science. "Oh, give us a few hundred years and maybe we will have firm, reliable design laws, but not yet. If only our field were as advanced as, say, bridge design, the world would be a better place."

Hah! The good news is that we actually do know a lot about how to design things appropriately for use by people. Moreover, most things designed for human use, especially electronic and computer devices, are so bad that it doesn't take much knowledge to make dramatic improvements. The problem is that the people who do these designs are often untrained as designers and insensitive to the needs of the people who use them. This insensitivity means that they never get around to consulting with the design community, with those of us who are sensitive and who have some knowledge of the design issues. Our design tools are not quite as good as the ones available for designing bridges or electrical circuits, but they are reasonable.

That's the good news. The bad news is that our standard measure of a well-developed, perfected class of design—the design of bridges, buildings, and other mechanical structures—isn't all that good.

Bridges, buildings, roadways, and roofs fail. Often without warning and without any immediately obvious cause. During the writing of this essay, the roof of a new building on a college campus in California collapsed. The college had to close down several other, similar buildings and launch a full-scale investigation.

Cargo doors fly open on airplanes, sucking out passengers in midair. Failures occur in mechanical design because, well, design is an evolutionary process, and each failure teaches us something that we hadn't thought of before. Today's accomplishments are built upon a history of past failures. Many of our most important design lessons have been discovered through careful systematic study of past failures.

Engineering design is unlike evolution, however, in that the lessons learned from failures can be applied immediately. Moreover, the new insights are directional, moving toward improvements. In evolution, if a variation in a species leads to degraded viability, that species dies out, but no permanent lesson is learned. The evolutionary process simply tries again, probably not repeating the same variation, but not necessarily choosing anything better on the next attempt. In fact, the lack of repetition is not by design but simply because each possible modification is so unlikely that it is very rare that the same path—whether good or bad—would ever get repeated.

In principle, design has certain virtues:

There is a specific goal.

The designer or inventor can guide the process.

A successful design can be redone, eliminating deficiencies, historical false paths, and unnecessary components. The designer can look back over a design and "clean it up."

Failures are informative, yielding clues about what went wrong and what ought to be changed.

All this means that, in principle, design is superior to evolution, because it is guided, because the designer follows engineering principles and can remember and benefit from previous successes and failures. That is the principle. In practice, design isn't that much different from evolution.

In the real world, there are huge pressures on the designer to get something done quickly and with less money ("we needed it yesterday, and we are far over budget"). As a result, there is less reflection and analysis of past successes and failures than you might think. Today, for example, we take the train, the airplane, and the boat as standard, safe means of transportation. We forget the history of their development. In the early days of each, travel was very risky. Steamships used to explode with great regularity, in part due to the lack of knowledge about the strength of the boilers, in part because captains used to tie the safety valves shut in order to get ever more speed out of their vessels (economic pressures, once again). The U.S. Congress finally launched an investigation, and the result was a series of scientific studies of steam engines that finally managed to increase their safety.

Railroads suffered explosions of their engines, plus derailments and collisions. The problems of scheduling multiple trains over single tracks led to a number of technical advances in business management, plus the development of standard means of determining time. We owe today's time zones, in part, to railroads. And perhaps to the thousands of passengers who were killed or injured.

Design in this case was like evolution, with repeated failures, and a slow, steady learning through those failures.

Why Evolution Is Not Design

Evolution is tinkering—what in computer parlance is called "hacking." Take something that exists, tinker with it in small, minor ways, and try it out—see whether the result succeeds or fails, where success or failure is measured by the arbitrary and capricious hand of fate. Does the new version lead to more or less progeny than the old? Actually, even the number of progeny is not the criterion—there is no criterion. Nobody is directing things, so nobody is looking or measuring or deciding.

Species are not manufactured products that get discarded when new technologies arise: Species live out or die out for complex reasons involving the interactions of numerous variables. What we think of as the "direction" of evolution or the functions of evolution are in the eyes of us, the beholders, not in the evolutionary process itself.

Why do new species emerge and old ones grow extinct? Note that even the most rapid evolutionary process occurs so slowly and gradually that if you were there, watching, you probably could not mark the time when a change took place. It is a bit like watching the sky to determine when it becomes dark. The dark part of night is very different from the bright part of day, but the boundary between them is a smooth, gradual transition.

The history of evolution is maintained because each new variant of a previous animal is simply the old one modified. This means the course of evolution is intractably linked to the earlier products. Moreover, because of the manner by which evolution works, a blind varia-tion on existing forms, the *structures* of evolution are irreversible, except by rare chance. Once an organism develops eyes or legs, tails or wings, those structures will remain even as their functions disappear. So it is that humans still have the remnants of a tail. *Functions* are reversible, which means that a bird can lose the ability to fly, but it will still keep its wings, as have the ostrich and the penguin. Fish that live in caves still have eyes, even though they are blind. Marine mam-mals such as dolphins, sea lions, and whales still breathe air and have all the characteristics of land mammals. Breathing air means they have to develop very special means to sleep, or else they might drown. Even today we carry with us the history of our forebears in various parts of our anatomy that do not work well for today's life:

> The appendix, that mysterious part of the body that seems to
> do little good and occasionally flares up to do harm.
> The spinal cord, a masterpiece of mechanics and wiring, but not
> well suited for the posture of modern humans, thereby the
> cause of much suffering among those of us past middle age.
> The digestive system, in earlier times adapted to a leaner diet—
> or perhaps a shorter life span. Because the modern "civi-
> lized" diet is high in sugars, salts, and fats, it does horrible
> things to the heart, lungs, and blood vessels.
> The immune system, a wonderful protecting device that can
> fail under the onslaught of modern substances. The result is
> allergies, self-attack (when the immune system thinks that
> normal body stuff is an invader), and other ailments of our
> time.

Are We Stuck with the Past?

Human design should be free of all of the difficulties of evolution, right? Nope. In many ways, human design is just as burdened with the structures of the past as is evolution. There are several reasons: The first has to do with the limited imagination of designers, the second with the heavy inertia of human organizations and standards, and the last with the "installed base" problem.

As we have already seen, designers are, well, only human. Most designers are not that creative. They look at what has gone before and copy it, making only whatever changes seem required for the current product (or to avoid being sued for copyright or patent infringement). Most designs borrow from previous designs because that is the only way people know how to build them. It is the rare breakthrough that produces a new way of doing things, and even then, the vast majority of designers will rush to copy the new method. This is not that much different from natural evolution.

The second problem is the huge inertia of society. Changes are resisted. Some change is resisted for cultural reasons—people see no need for it: "What was good for our parents is good enough for us." Some change is resisted because it carries a high cost in learning or retraining, perhaps a loss of jobs. Some change is resisted because it entails new ways of doing things, which means unknown difficulties and pressures. Unions resist changes that appear to be instituted at the expense of the worker. Management resists changes for much the same reasons—if their jobs appear to be in jeopardy, or if new skills and work seem to be required. All are very understandable reasons, and often the fears are justifiable.

National standards are essential to ensure that everyone can use the same products, and that installations are performed safely. But standards can hold back progress as much as they assist. The international standards for radio transmission are archaic, wasting large amounts of bandwidth to produce low fidelity signals. Modern methods of digital encoding and distributed, modular transmissions could increase both the number of available channels and the sound quality. Nonetheless, today's standards persist because the amount of effort to get international agreement for a reallocation of radio frequencies and a change in broadcasting standards is more than most people can contemplate. It will take generations (and in fact, already has). The same is true for television.

The last problem is that of the installed base. The term refers to the large number of existing devices that would be made obsolete by any changes. One reason for the slow changes in international standards is the installed base problem: If we were to change the method of sending radio signals, then the hundreds of millions of AM and FM

114

radios in the world might all become obsolete, and worthless. This is more than most nations wish to inflict upon their citizens.

So just as biological evolution is wedded to its past, industrial design has its own tyranny of the past. Consider the list:

Typewriter keyboards: What do you do with today's typewriter keyboard, all the world over derived from the original American standard "qwerty" layout. This is known to be an inferior keyboard: better arrangements exist. But there are so many millions of keyboards installed in the world that it now seems quite impossible to change without enormous cost.

Men's and women's fashions: Are men really stuck with the tie and women with high heels, forever?

Nonphonetic spelling: What about the quaint spelling of the English language? Or French?

Multiple languages: Will we forever have umpteen thousand languages in the world?

Feet and ounces instead of meters and grams: Is the United States really stuck with archaic English units of measurement, while the rest of the world has switched to the more logical, easier-to-learn, easier-to-use metric system? Even the English no longer use English units.

Old-fashioned television standards: For how long will those of us in the United States be stuck with our old-fashioned standard for color television, NTSC. The initials NTSC stand for "National Television Standards Committee" but those in the TV industry usually call it "Never The Same Color" because of the great difficulties of transmitting and receiving accurate colors. Next time you watch a talk show, notice how the color of an actor's face changes when the picture is switched from one point of view to another (which indicates that the camera changed). This shouldn't happen.

The NTSC standard was first adopted in the 1930s for black and white, and then modified in the 1940s for color, but with the restriction that it still had to be compatible with the 1930's black and white. Thus these standards are over a half-century old. They were created in ignorance of all the improvements in knowledge about sound, picture, and information transmission that have taken place since they went into effect. They haven't been changed because of the expense for all existing set owners, and television recording and transmitting studios. We will get rid of NTSC, but the fight to do so is taking decades.

Differing standards for television across the world: Today the video signal used in Europe differs from that in the United States or Japan. The standards are different, and one country's set can't display another country's signal. The world has three major ways of encoding the television signal: Think of them as three different species. In this age of worldwide communication, wouldn't it be nice to have but a single standard? And now that the world is indeed moving to new standards for high-definition television, wouldn't it be nice if everyone would agree on the same one? Yup, but don't count on it. So far there is no sign of the cooperation that would be necessary for such agreement. Each part of the world seems to want its own system.

In other words, design is not free of its own historical pressures. Design is frequently copying from others, borrowing upon what has come before. Even if the copying is not deliberate, it is there, because we can do only what we know, and knowledge is cumulative, building upon what has come before. So, design is more often modification than innovation, following the history that came before, even if it is no longer relevant, even if we no longer know what that history meant. Is design so very different from evolution? I think not.

11

Turn Signals Are the Facial Expressions of Automobiles

NATURE produces a varied assortment of creatures. One that has long fascinated me is the red-tailed baboon. You know, the one with the, umm, red-colored rear? All that color and display, but for what purpose? Well, looking at the rear end of some automobiles reminds me of looking at the rear end of some baboons.

Social cooperation requires signals, ways of letting others know our actions and intentions. Moreover, it is useful to know the reactions of others to our actions: How do others perceive them? The most powerful method of signaling, of course, is through language. Emotions, especially the outward signaling of emotions, play equally important roles. Emotional and facial expressions are simple signal systems that allow us to communicate to others our own internal states. In fact, emotions can act as a communication medium within an individual, helping bridge the gap between internal, subconscious states and conscious ones.

As I study the interaction of people with technology, I am not happy with what I see. In some sense, you might say, my goal is to socialize technology. Right now, technology lacks social graces. The machine sits there, placid, demanding. It tends to interact only in order to demand attention, not to communicate, not to interact gracefully. People and social animals have evolved a wide range of signaling systems, the better to make their interactions pleasant and productive. One way to understand the deficiencies of today's technologies and to see how they might improve is to examine the route that natural evolution has taken. You know the old saying that history repeats itself,

117

that those who fail to study the lessons of history are doomed to repeat its failures? Well, I think the analogous statement applies to evolution and technology: Those who are unaware of the lessons of biological evolution are doomed to repeat its failures.

Evolution works its way slowly, ponderously. Even those who believe that it progresses in rapid jumps think that these jumps take tens of thousands of years. By the standards of an individual human, even the most rapid evolutionary changes are too slow to have any impact on the individual. But the study of evolution might aid us as we design artificial devices, enabling us to profit from evolution's experiences, letting us accomplish in years what has taken millennia for evolution. Over time, evolution tries out a wide variety of methods to ensure survival, some that modify the animal or plant, others that modify behavior, and still others that affect the cooperative, interacting nature of social structures. The female baboon's red rear is one result of this process, as are other signaling systems, such as the calls and cries of animals and the gestures, facial expressions, and the speech of humans. We can learn from evolution's successes.

Animal Signals

Animals require a number of different signaling systems to communicate internal states both within themselves and to others. Plants signal their maturation and ripening through colorful displays that attract the attention of insects and birds, the better to pollinate and propagate their seeds. These signals need not be consciously given or received to be effective. All that matters is that there be some perceivable change of state that other organisms can make use of. Thus the presence of snow, ice, and objects waving in the wind signal a state of weather without any conscious volition on the part of the atmosphere. But the signals, nonetheless, are valuable ones.

The red rear of the baboon is certainly not a conscious signal, and for that matter, neither are many of our facial expressions. Facial expressions originated as side effects of the facial muscles as they prepared the mouth, lips, and teeth for activity. But as they were perceived and used by other animals, they began to evolve toward a

118

symbolic, meaningful role, so much so that today many of our facial expressions are voluntary, conscious, and deliberate.

Facial and body expressions have evolved because they serve a useful purpose. The signaling of intentions and internal states among animals works to their advantage. The importance of facial expressions was recognized quite early in the study of animal evolution. Charles Darwin devoted an entire book to their study—*Expression of the Emotions in Man and Animals*. Emotional expressions act as a side channel of normal communication, outside of and without interference to the spoken language. They offer a commentary, and when this information channel is lost, ambiguity and difficulty in interpretation often results.

Social interaction requires a very different set of behavior patterns than does solitary action. Successful social interaction means cooperation, joint planning, troubleshooting, play, rivalry, competition, and comradeship. It means social honesty as well as deception. If animals—including people—are to form functional social groups, they must develop means of communication, of synchronizing actions, of cooperation, and occasionally, of deceit. Social interaction was the driving force for the evolutionary development of social signaling devices. Facial expressions, colorful plumages, the red tail of the baboon—all are social signals useful to the communication, interaction, and protection of animals.

There are many controversies surrounding the development of human intelligence. The traditional views link the development of intelligence to language or tools, or perhaps to the need to be flexible and innovative in dealing with a changing, complex environment. There is much to be said in favor of all these possibilities. Recently, however, a new suggestion has emerged: Higher forms of intelligence result from the need to handle the problems of social interaction. This is an attractive notion, for social interaction requires numerous talents and abilities, including the ability to let selected participants in a social group know the intentions and beliefs of others. Human intelligence almost certainly did not result from any single factor. It is most likely the result of multiple forces acting over long durations, but social interaction seems like a good candidate to be one of the primary forces.

One of the important aspects of intelligence is the ability to communicate. There are actually two levels of communication that we need to be concerned about: one is internal, between the body and the conscious mind; the other is external, among animals and people. Internal signals are very important for the individual. Thus the body informs the mind when it is experiencing heat or cold, injury or chemical imbalance. The results are subjective feelings of warmth or chill, pain or discomfort. Emotions also serve as internal signals: from happiness, pleasure, and love, to sadness, anger, envy, and dissatisfaction. Emotions are complex mixtures of biological and mental states, the neurochemistry of the brain interacting with the information processing of events and expectations.

How many times have you been in situations where your mind tells you one thing and your body another? How often have you delayed a decision, saying "I want to think about it," but really meaning, "I want to see how I feel."

Some people believe that emotions are a vestige of evolution, a type of "animal" behavior that the human race will eventually outgrow. Well, not really. If you look at the role emotions have played in evolutionary history, it would appear that the more sophisticated the animal, the greater the role played by emotions. We are the most sophisticated of all, and thereby the most emotional.

Emotions speak to our cognitive minds, sometimes telling our conscious selves things we would rather not know. Emotions also trigger body changes and facial expressions that signal others. Moreover, because other people in our social circle are apt to share similar backgrounds, the same basic knowledge, and the same biological mechanisms, we can assume that things that make us happy or angry are apt to make them happy or angry as well. Taking the point of view of another would be aided if only we could read the mind of the other, and this is where external signals come in.

How do we tell what others are thinking? We do so through a variety of techniques, with varying degrees of accuracy. Facial expressions, gesture, and body position act as cues to a person's internal states. We often call these things body language, the name indicating the communicative role. Body language makes visible another's internal state. The blush of the cheeks, the grimace, the frown, and the smile all act as readily perceivable external signals of a person's internal state, making visible to observers what would otherwise be difficult or impossible to determine.

Of course, facial expressions can deceive. First of all, they are subtle, a rich interplay of complex musculature, facial features, and coloration. Body language is even more complex. Not everyone can read the signals, and for that matter, not all scientists are convinced that the signals are there. When I cross my legs, am I sending a subtle message or am I simply trying to get comfortable?

The study of human emotions, of course, is a complex topic, one that has occupied psychologists for years. We are learning a lot about the neurochemistry of emotions. We know that there is a complex interplay between the neurological state of arousal and the cognitive interpretation of the state. One long-standing argument is over the ordering: Which comes first, the emotional state or the interpretation? One side of the argument says that, essentially, you first notice your body's emotional state and then interpret it: "Oh, oh. My heart is pounding. I feel tense. I'm sweating—I must be afraid." The other side says no, the interpretation leads to the state: "I have to give a really important presentation tomorrow. I'm not ready. I can imagine everyone looking at me, their eyes registering skepticism. I'm afraid. There, see: My heart is pounding. I feel tense. I'm sweating." Both sides of the debate have merit, which means that the true story is probably a combination of these two differing views. To the onlooker, however, it doesn't matter how the emotional state has been aroused. The facial expressions and other body signs are external indicators of the resultant state.

To speak only of the body signaling the conscious mind is a great simplification. Nonetheless, simplification is a useful starting point, useful and scientifically reasonable. Moreover, one person's interpretation of another's mental state really does depend upon this information. Of course, the visible signs of emotion are often ambiguous, incomplete, and misleading, but so too are our judgments. Our interpretations of another's feelings and beliefs are not very accurate and often are misleading.

Animal Deceit

The fact that facial expressions and spoken language can be used to communicate internal states is useful, both for conveying accurate information and for the deliberate deception of others. I can appear happy when in fact I am not, feign sadness for an event that pleases me, or disguise my knowledge in a variety of ways. The study of deceit in animals is a powerful way to get clues about their level of intelligence, clues to the working of evolution. As a result, there is an ever-increasing number of studies concentrating upon the evolution of deceit, the false communication of knowledge, intentions, and actions. Deceit seems a natural property of higher animals, and not just to protect the young from predators. Chimpanzees, for example, use deceit to avoid sharing food, to curry favor with higher-ranked animals, and to obtain sexual favors behind the backs of disapproving higher-ranked males. These kinds of deceits may very well require more intelligence and cleverness than truthful communication.

Deceit is a necessary part of civilized life—not the evil deceit where one person seeks to benefit at the expense of another, but the polite deceit of social interaction. The polite "thank you" when someone presents an unwanted gift, or the polite acknowledgments when someone else's well-intentioned efforts go awry at your expense. Social interactions require falsehoods to maintain themselves, and little benefit would result were every negative thought transmitted to others. Mind reading may seem like a desirable trait, but it would often backfire, causing grief where none was intended or desired. All societies and cultures have developed social codes that govern interactions and mask true feelings and beliefs under a cloud of cordiality or, at least, civility. The elaborate honorifics and "speech acts" of society exist for good reason.

Lies and deceits have their place in the world. Social interaction would be less pleasant if the truth were always told. Casting blame on the other person is an excellent administrative policy. When I need to make an unpopular decision, I sometimes blame my administration— "I'm sorry, but my Dean won't let me do that"—checking first with my Dean, of course, to make sure that my stand is understood and approved of. In turn, I advise the people who work for me to use the same tactic, to tell others that they are sorry, but their boss won't allow it. In fact, it works in both directions. When my Dean wants to resist an order from above, he can call the department chairs that report to him and get us all to allow him to say, "Sorry, but my departments won't agree."

Social interaction is complex. We are the most social of all animals, and we have evolved elaborate schemes for interactions, schemes that allow us to coexist and cooperate with friends and to resist the pressures of enemies. The act of deceit is complex, and most animals are incapable of it, not because they are more honest but because their brains are inadequate. Only the most sophisticated of beings can lie and cheat, and get away with it.

There is, by now, a large research literature on the abilities of monkeys and other animals to deceive. One of the marks of evolutionary development is the evolution of "social artifacts," the ability to use social interactions and strategies for cooperation and for deceit, all to the betterment of the social group. It was a relatively late development in evolutionary history. Among the animals that are capable of practicing social cooperation, the human is superior at forming tight social bonds as well as in using the social deceits necessary for cohesive social structures.

Only the most advanced of primates, the ape family (which includes the gorilla and the chimpanzee), seems capable of true deceit. Monkeys sometimes try, but it is a bit too much for their minds to manage. Look at this description of an African monkey, the vervet, trying to deceive his rivals:

> *A male vervet, Kitui, gave leopard alarms when challenged by a rival male, causing the other male to flee up a tree.*

That is real deception: To get rid of his rival, Kitui called the monkey equivalent of "Fire! Fire!" The problem is, the monkey couldn't quite pull it off.

> *However, to reinforce his point, Kitui descended from his own tree and walked across the open ground toward his rival, still calling the equivalent of "Run for the trees." . . .*

[This is like] a human three-year-old who with crumbs all over his face denies having raided the cookie jar.

Notice that the vervet is intelligent enough to give a false signal in order to scare away a rival but not intelligent enough to act out the entire behavior. It doesn't really matter for vervets because the rival male isn't sophisticated enough to see through the ruse, to realize that Kitui's behavior indicates that the call is a fake. Young children will make the same type of error that Kitui made, but parents are quite capable of seeing through the deception. So, for this kind of intelligent behavior, a vervet is acting somewhat like a young child.

Notice what Kitui would need to realize in order to do the ruse correctly: He would have to know not only that the leopard alarm would cause the rival to flee but that his own behavior must be in accord with the falsely announced state. The animal needs to know not only what a signal means but how its signal and its own behavior will be interpreted by others, and then to understand that one can contradict the other and that others can perceive this contradiction.

A chimp wouldn't be fooled by Kitui. Thus a chimp has been seen using his fingers to readjust his mouth in order to hide a grin before turning to bluff a rival. This behavior shows that the chimp is aware of his own facial expression, aware that it is visible to rivals, and probably aware of how the rival will interpret the expression. I have seen the same behavior in adult humans.

Artificial Devices and Artificial Evolution

As we construct artificial devices with ever more power, ever more intelligence, perhaps we will have to make them mimic natural evolution. Technology slowly evolves, not in the same way as the natural evolution of life but through the artificial evolution of design. But in many ways, the evolution of machines is driven by the same pressures that drive the evolution of life. Modifications that enhance performance and allow the organism or machine to survive and to compete in the world will survive; those that do not will disappear. Slowly, designers will add signals and warnings, self-assessments and communication devices, providing the artificial equivalents of emotions, facial expressions, and social interaction.

Natural evolution combined with cultural conventions determine the nature and interpretation of the facial expressions of people.
Machines pose an interesting problem, for they are artificial devices,
manufactured by people. It is the rare machine that works entirely
alone, isolated from other machines and from people. Machines have
to be started, stopped, monitored, adjusted, and maintained by people.
Many require considerable control by humans. Machines are social
devices, for their manufacture results from interaction with people.
As a result, some of the same pressures that gave rise to facial and
emotional expressions in animals apply to machines as well, except
that here, the signals have to be designed, deliberately constructed,
and integrated into the machine. The lights and sounds of an automobile play a role analogous to the facial expressions of animals, communicating the internal state of the vehicle to other vehicles in its social
group.

With animals and people, we saw that there were two different
forms of signals: internal and external. The same is true for machines,
except here, we have to readjust our idea of the basic "unit" of analysis. For a person or animal, I distinguished between the body and the
mind. Internal signals informed the conscious mind of the subconscious information: body states such as hunger, fatigue, and comfort
as well as emotional states such as fear, joy, or anger. With a machine,
there is no such thing as the conscious mind—but there is the user.

Machines can be regarded as symbiotic units consisting of a machine and a person. Thus a copying machine forms a functional unit
only when combined with a person: person + copying machine. Similarly, the automobile by itself is not functional: The critical unit is
driver + car. The person has the same relationship to the machine as
consciousness has to the body: a supervising element that watches
over and maintains the system, even as the system—human or machine—sometimes operates relatively autonomously. This is a dangerous metaphor to pursue deeply, for it fails in all sorts of ways, but it
does have useful characteristics with regard to the way in which the
machine ought to interact with people.

Just as animals have two levels of signals, internal and external,
there are two levels of interaction of this person + machine unit—one
internal, the other external. Internal signals in a person tell of body

states, but in the person + machine unit, they tell the person about the internal states of the machine. A machine signaling a person isn't really the same as a body conveying internal information to consciousness. The machine is performing a form of social communication between it and an outside agency, the person who is using or maintaining it. So the signals between machine and human have to be a combination of external, social signals and internal ones: internal to the unit of person + machine but external in the sense that the machine and the person are separate entities.

One of the special kinds of signals that this relationship requires is feedback about the operation itself. It is difficult to use a machine that does not provide feedback to the user. Mechanical devices tend to do this through their construction. A pair of scissors feels firm or loose: Its blades snip-snap through the air with a pleasant sound, or scrape, moved only with great force. Or they might wobble, providing a sense of insecurity. A good knife provides feedback through its balance and feel as it cuts. Mechanical devices are often visible and audible, conveying considerable information about their operation, even to those who know nothing of mechanics. The designers do not have to provide feedback to the users. The very nature of the machine guarantees that.

Not so with electronic devices. Electronic devices work quietly and smoothly, invisible and inaudible. At most, one might get a hum, buzz, or crackling sound resulting from components that vibrate with changes in magnetic fields or from heating and cooling. These sounds, however, are peripheral to the operation and seldom convey useful information. But more important to the user is that electronic systems deal with information, not mechanical movements. Information is a commodity that exists conceptually, not physically. It occupies no space, makes no sound.

One of the reasons that modern technology is so difficult to use is because of this silent, invisible operation. The videocassette recorder, the digital watch, and the microwave oven—none is inherently complicated. The problem for us is their lack of communication. They fail to interact gracefully. They demand attention and services, but without reciprocating, without providing sufficient background and context. There is little or no feedback.

There are many reasons to need feedback about the state of a system, reasons dealing with our own need for knowledge and reassurance. This kind of feedback is essential in normal social intercourse. The spoken "hmmm" or the nodding of the head by the listener to a conversation assures the speaker that the message is being received. The feel of the screw's resistance to my turning of the screwdriver provides useful feedback about the success of my operations. Feedback is a necessary part of all interaction, whether with people or technology, but it is more absent than present in today's information-based technology. If our information-based technologies are to become socialized members of society, interacting with and supporting the activities of people, then they have to be able to interact with us on our terms, not on theirs.

Our most modern technologies are social isolates. Today's technology provides us with ever-more complex machines, devices that can work at a distance or through nonmechanical components. Humans are often unaware of their presence, unaware of their internal states. The modern information-processing machine fits the stereotype of an antisocial, technological nerd. It works efficiently, quietly, and autonomously, and it prefers to avoid interactions with the people around it.

Just as it is valuable for us to know of our own internal states, the better to manage our own existence, designers of mechanical devices need to signal the internal states of their machines, the better to keep them maintained and functioning. The hunger and thirst of animals translate to the energy supply of machines, perhaps specified as the fuel level or the state of the battery. Is the machinery too hot, too cold? Lubricated properly? In appropriate adjustment? These are the things that the person maintaining the machine needs to know. Such information is provided naturally by the human body, but it must be provided artificially for our artifacts.

This internal information is provided in a number of ways. Sometimes it is not given explicitly but rather is stated as rules ("lubricate every six months"). Sometimes it is assumed that the user will notice and repair deficiencies as they appear ("tighten connections as needed"). More complex machines require indicators of their internal states, and these are provided by lights and gauges, by instrument panels. The instrument panel of an engine shows critical aspects of its internal state, thereby allowing the driver to control it safely and efficiently, perhaps much as the hard-driving coach carefully monitors the emotional responses of the players, attempting to push them hard enough to do some good, but not so hard as to be destructive.

The instrument panel of the automobile allows an internal communication within the driver + car unit. It is mostly self-centered, communicating information from the machine to the driver. Most instrument panels are like that: Lights tell us whether a device is turned on, meters and other indicators tell us of the current state, buzzers and alarms tell us when something is wrong and needs immediate attention. No social protocols, no etiquette. No checking to see whether we are busy at some other activity, usually not even a check to see if other alarms or warnings are also active. As a result, when there are serious difficulties, all the alarms and warnings scream in their self-centered way, the simultaneous array of lights and sounds impeding intelligent actions by the operators of the system. In places that have large control panels, such as industrial control rooms, commercial airplanes, and even the hospital operating room, the first act of the human operators is to shut off the alarms so that they can concentrate upon the problem. Unfortunately, the machines have no way of learning from the experience—you can't spank them and send them to bed, nor is there the equivalent of a note to the parent. As a result, when trouble next strikes, the same rowdy behavior reappears.

Social issues are even more serious when we consider socially interacting units of machines. The prototypical example is driving, where the driver + car unit interacts with large numbers of other similar units. Here is where we have a need for external signals, signals that communicate with the other units sharing the road, signals that allow others to know just what actions are being performed, and in many cases, what actions are intended. There are times when it is necessary to know what is on another's mind.

The same technology that makes modern transportation so efficient would kill us without rules of social behavior. Thus vehicles are restricted to certain locations. Similar directions of travel are put into the same corridors with some separation between those going in other directions. For safety, order and regulation are essential, even in societies that normally shun order and regulation.

With automobiles, we use traffic lights and signs to indicate who may go and who must stay, what can be done or not done, and who has precedence over another. And we use turn signals and brake lights to tell others of our actions and intentions. In the case of brake lights, we signal actions as we carry them out. In the case of turn signals, we signal our intentions before we actually commit them into action. In either case, we allow others to know our future actions so that we can ensure that there is no conflict.

The brake lights of the automobile serve no function for the operator. Rather, they are a way to communicate with other drivers. The brake light means that the brakes are applied, which the other driver interprets to mean that the car is slowing down. Moreover, the other driver will usually search for a reason, some explanation of the brake lights. This means that the lights can serve a valuable communication purpose. On a long, normally noncongested highway, if the car in front of you applies its brake lights, it usually signals some unexpected obstruction or danger on the road, and it is usually wise to slow down and be more alert.

And here is where intentions come in. Social interaction is enhanced when the participants know not only what is happening at the moment but what will happen next. Of all the signals of the automobile, only the turn signals announce intentions.

Intentions are tricky, for they play many roles in social interaction, some obvious and necessary, some subtle and devious. In games we often signal intentions in order to deceive. Of course, our opponent does the same, and interprets our signals knowing full well they may be deceptive. Thus starts the elaborate ruse and counterruse, where we try to determine how other people are reading our minds. Suppose that in a game I want to kick the ball to the right. I could first pretend to kick the ball to the left, but I know that they will know that this is a pretense. But if I pretend to kick to the right, they will expect me to kick to the left, unless they know that I will think that, in which case they may realize I really intend to kick to the right. So I decide to fake a kick to the right, and then run. Except that when the actual time comes, things may happen so fast that none of the plans can be executed. In games and war our signals are more often false than true.

Imagine doing this in traffic: signaling a left turn, hoping that this will open up a hole in traffic that will let you dart to the right. I once got a driver's license in Mexico City, where aggression was the rule. But even there, intentions had to be signaled honestly. Above all, it was essential to avoid eye contact with other drivers. In the traffic circles of the city, the trick was to avoid letting the other drivers see that you had seen them. Once the other drivers knew that you knew they were there, they would proceed at high speed around the circle, completely ignoring your presence, because they knew that you knew that they were there, so they expected you to stop or slow down. And you had to, or be killed. On the other hand, if you could manage to avoid letting them see you see them, you could proceed with impunity, because now it was their responsibility to avoid you. If you collided, it couldn't have been your fault, because after all, you hadn't seen them.

Most places in the United States don't let you get away with such games. In my community in southern California, for example, fault and blame are mechanically assigned according to strict orders of precedence. The rules of the road determine whose responsibility it is to avoid accidents. Thus, at intersections, the automobile on the right has the right of way, and all the eye contact in the world won't change that.

In Mexico, there were other ways of signaling intentions. Thus, if two cars were approaching a narrow, one-lane bridge from different directions, the car that first flashed its lights thereby announced that it was coming through, so the second car had better yield. The flashing headlight was to be interpreted as, "I got here first, so keep out of my way." As long as everyone understands such signals, they work fine.

The problem is that other cultures can completely reverse the meaning of the signals. In Mexico, one wins by aggression. In Britain, one wins by politeness and consideration. So in Britain, in a similar situation, the car that flashes its lights first is signaling, "I see you, please go ahead and I will wait." Imagine what happens when a Mexican driver encounters a British driver.

Drivers of automobiles get pretty good at reading the intentions of others. Brake lights and turn signals offer a formal, mechanized set of signals, eye contact another. Headlights also serve a valuable communication purpose, with the flashing lights conveying many different messages, depending upon the circumstances. Horn blasts and hand signals are also used. Basically, any part of the automobile that other drivers will recognize as being under the control of the driver can be used to signal something.

Notice that not all signals are truthful. Deceit exists on the highways just as much as it does in other social endeavors. The impatient driver can try a variety of tactics to gain open roadway, although the flashing of lights or blaring of the horn is the most common, most direct method. Some signals are ambiguous or confusing, such as the turn signal that continues for block after block, either signaling the eventual desire to turn or simply a remnant from a previous turn.

Turn signals are peculiar devices, neither human nor artificial. They are really not a way for the automobile to communicate with people. Turn signals are simply an aid to normal human-human communication. Instead of shouting or pointing, we simply flip a small lever here, resulting in visible flashing lights there, outside the automobile. Even so, turn signals are an important start toward the graceful interaction of people and machines.

The Graceful Interaction of People and Machines

Human social interaction has developed a rich assortment of methods to ensure social harmony. Every culture has developed means of maintaining politeness and courtesy, of communicating needs without offending. If machines are to interact successfully, they too must follow these conventions.

Designers of machines usually provide the critical signals of the machine's internal state, for they know that maintenance is essential to operation. But then they often stop, failing to take into account the needs of the user of the device. As a result, the machines are still stuck in the asocial world of isolated devices. Worse, they have no manners. If machines operate in isolation with no need for interaction with people or other machines, then the lack of social graces and feedback about their internal states can be excused. But when machines are intended to operate with people, then the lack of socialization can lead to difficulties. Think of the telephone, continually intruding upon conversations, insensitive to the ongoing activities, forcing interruptions through its demanding ring whether the time is convenient or not. So it is with most machines, shunning interaction except to demand attention. We call such behavior in people "spoiled," "arrogant," or "insensitive," but somehow we have accepted it from our machines.

Social cooperation requires more than letting others know your actions and intentions. It is also necessary to know how the others have received your communication: Did they understand? Do they approve? Will they abide by it? When I talk with someone, I need to know how they are responding. Are they interested or bored? Do they understand or are they confused? When people engage in joint activities, they need to agree upon the division of activities in advance in order to be able to synchronize and coordinate their efforts, to avoid conflict. And during the activities, knowledge of the other person's actions is important, if only to know that the person is still interacting. It really is essential to get some feedback, if only to hear the "hmmm" from others. Otherwise how do we know they are attending? How do we know whether or not they are even alive?

We do have machines that are showing some of the first, early signs of social graces. Some can guide expectations, and even question actions. Thus, in the word processing system I am using to write this chapter, if I try to do a complex operation, I might be warned:

> *The current action cannot be undone: Do you wish to proceed?*

Or if I try to move a file from one location in the computer storage system to another, I am sometimes warned:

> *An item with the same name already exists in this location.*
> *Do you want to replace it with the one you're moving?*

These are early signs of social maturity: polite, meaningful concern about the possible effects of the operation I requested be performed. Yes, some artificial devices show the early signs of social responsibility, displaying their internal states for others to see, sometimes assessing the impacts of their own actions and warning others of them. The interaction is primitive, however, and often not as effective as one might expect from human colleagues. The subtleties and richness of natural emotions and natural attentiveness to social interaction are missing. But even so, the first glimpses of artificial systems that exhibit cooperative, social behavior are appearing.

Perhaps the simplest form of social cooperation among artificial systems is the "handshaking" protocols of communicating systems, invisible to the normal human user but essential nonetheless. Handshaking is, of course, a human custom with a long cultural evolution. Today, shaking hands is part of the ritual by which people meeting for social or business purposes introduce themselves and get set for conversation or business. With machines, the term "handshaking" has been reserved for the initial steps of a communication protocol in which all the devices determine that they are connected properly, that their messages are in synchronization, and that they are directed to the correct recipients and in the correct format. It is easy to eavesdrop on these proceedings by listening to the first stages of a telephone connection of computer to computer or facsimile machine to facsimile machine. Better yet, listen as a computer tries to talk to a fax, or even to a person. Then the handshaking fails, as the automatic system tries this protocol and that one before quitting and gracelessly hanging up the telephone.

It is a sign of our technological era, of course, that elaborate handshaking protocols and other social niceties have been developed to handle the interactions among machines, but that no such civility seems to have become standard for the interactions between machines and people.

Human emotions, facial expressions, and social interaction have evolved over millions of years. We have had time to do things slowly, to work things out with care. Even so, there are occasional mismatches as people fail to understand one another, fail to cooperate.

What will happen with our machines? In principle, artificial evolution can proceed much more rapidly than can biological evolution. Artificial evolution can take advantage of knowledge and experience, but so far there is little evidence of attention to these.

I fear that the rush to autonomous machines is proceeding too rapidly. Our machines are barely social now. They are still at an early stage of development, still primarily self-centered, still focused on their own needs and not those of their operators. What will happen when they are given more power, more authority? How can we shape the evolution of machines so that they become more humane, more in line with human needs and values?

Mind reading is an essential activity for social communication. If I am to interact with machines in a constructive manner, then I need to be able to do the equivalent of reading the mind of the machine. Machines don't have minds, but they do have internal states. More and more, they are able to have goals and plans, expectations and even desires. The interactions will be smoother and more friendly if I can know these things. Facial expressions are rich and varied, a lot richer in information content than a few lights or sounds or meters. They reflect many subtle variations of mood. The blush that affects facial color, plus body position, and the sound of the voice, all give subtle indications of the underlying mood. Would that our machines were so sophisticated.

Just as people need to communicate acts, intentions, and emotional states, to give continual feedback and evidence of expected actions and outcomes, so too will machines have to interact more fully, more completely, to provide the same kind of information. Will we have to repeat the whole ensemble of human emotional and facial expressions in our artificial devices? Yes, I think so. The history of technology might very well have to repeat the history of the social development of humans: Technology recapitulates phylogeny.

12

Book Jackets and Science

I STARTED out to write an essay on the shelf space taken up by book jackets, but my investigations turned it into a case study of the workings of science. Scientists are very conservative. Nobody trusts anybody, which is how we make so much progress.

My scientific conservatism led me down an interesting path. I started out with a question: Do book jackets waste space? I knew that most people would say "of course not—they're too thin to make a difference." Henry Petroski, an engineer and author of several influential books, decided to find out. He measured the thickness of book jackets, and concluded that they waste the space of one book for every forty. So a library of one million books would waste the space of 25,000 books. Wow. Impressive. Quick, discard those book jackets.

But wait. Can this be right? It violates common sense. Now, violation of common sense is neither unusual nor undesirable. Common sense is frequently wrong. In part, the whole reason for science is to let us go beyond what common sense tells us. But still, whenever anything doesn't feel right, it is worthy of a second look.

Do we save so much space with a tiny, thin book jacket? Well, maybe. Let's check Petroski's figures. The first rule in science, by the way, is to check the other person's figures. Scientists are skeptical folks, never believing anything. That's why they are so very conservative by nature. Lots of people think of scientists as radical thinkers, always trying the outrageous. Nope. If anything, scientists distrust anything that differs from what has come before. So, first they check the data, or repeat the experiment. If the results can't be repeated (the scientific term is "replicated"), then they don't waste any more time.

If they can't repeat them, then there isn't anything to talk about—except why they can't repeat them.

So, let's start by repeating the results. For his analysis, Petroski used a micrometer and actually measured the thickness of a book jacket. I don't own a micrometer, but there is no need for one. We can approximate the result without any measurement. The book jacket has four layers including the inner flaps, so it is approximately the same thickness as four pages. Therefore the book jackets for forty books do indeed waste the space of a 160-page book. If a library of one million books got rid of one million book jackets, it would save space for another four million pages on its shelves. Hmm. So, even our informal way of measuring the size comes up with a similar answer: Getting rid of book jackets saves an enormous amount of space. Can this be right?

Now we come to the second rule of science. I told you scientists are skeptical. First, we are skeptical about the numbers. If they are OK—as they are in this case—then we are skeptical about the conclusion. So, can it be right that book jackets waste enormous amounts of space? Maybe not. There is nothing wrong with the arithmetic, but maybe there is more to the space issue than arithmetic: I was trained as a psychologist, the science of mental behavior, so let me think about the behavior, not the numbers.

Wait a minute, the space-saving argument works only if we assume that all one million books are on one shelf, so that the space savings add up. In fact, most book shelves are limited in size, and any extra space on one shelf does not translate into extra space on another. That is, if two shelves each have enough extra space for half a book, that space does not provide room for an extra book—there is no way you can actually put that extra book into those two half-spaces.

So how much space do we save in a library of one million books with covers if we take off the covers? It all depends upon how long the shelves are and how wide the books. The answer will range anywhere between no savings in space at all to savings for 25,000 books (Petroski's answer).

How could there be no space for more books? If each shelf were less than 1 meter in length (less than 40 inches) and each book at least 25 millimeters wide (at least one inch), then no space would be saved, for although there would now be almost an extra inch of space on each of the more than 25,000 shelves of this million-book library, no book could fit in the extra space on a single shelf.

What are the real numbers? I don't know. But the point is that the answer can vary across a very wide range of values. The actual answer for any real library is bound to be somewhere in between: neither a large value nor a zero value.

You see why it pays to be skeptical. In science, we don't trust anyone, even our best friends. But by always testing, confirming, and reconfirming the data, then by looking for alternative explanations, any new idea is put to a rigorous test and retest. In the end, the results are robust, scrutinized by numerous scientists around the world.

But the experimental scientists do not stop here. So far, what we have is theoretical science, a discussion between an engineer and a behavioral scientist, neither of whom actually works in a library. Things in the real world often work differently than theory proposes. Many other factors intervene, and the clean, pure predictions may not come to pass. The true test is in the application, which might even be the third rule of science: Check out the results in practice. So off I went to check with some librarians.

To summarize a long and fascinating set of discussions: Librarians don't concern themselves with the issue. Yes, they are very concerned about space, but that issue does not enter into the decision about whether or not to keep the jackets. Two major factors affect the decisions about jackets: First, whether there is information on them that their readers want; second, whether the jackets are durable enough to withstand hard usage. If a book jacket contains relevant information, either about the book or the author, librarians believe that their patrons would be most upset if the jacket were not on the book. But because most jackets are not very durable, the ones that are kept are usually laminated in plastic to safeguard them by increasing their strength and protecting them from dirt and abrasion. Of course, this make the books even thicker.

There are differences among libraries. University reference libraries don't usually worry about jackets; most of their books don't come with them anyway. Undergraduate libraries, on the other hand, do keep the jackets, for such libraries deal with more popular material and their readers expect to see jackets. The same dichotomy of views is found in public libraries. Their research sections are not apt to keep jackets, but the more popular sections do, again because it is perceived that jackets are demanded by their patrons. After all, jackets are often what attract a reader to a book in the first place.

So although libraries are very much interested in saving space, they place more importance on information content and the reader's interests. Moreover, librarians worry a lot about the durability of the books in their collections, so they often make books thicker in the interest of making them stronger. Book jackets get an added layer of plastic over them; paperback books are often rebound in hard cover to protect them.

So all those calculations were solely of academic interest. Librarians save the jackets. Then again, as I was finishing one lengthy conversation, the librarian said to me: "Would you give me those numbers again? You would save space for 25,000 books? Four million pages? That's really interesting."

Ah yes, one important role of theory is to be able to point out aspects of a situation that are not obvious without it.

The Difference Between Continuous and Discrete Science

One last puzzle bothered me. Why did Petroski do the simple space-saving computation as if space did accumulate? Why did he not realize that the discrete nature of bookshelves blocked this accumulation of space? Maybe the difference is that he, the civil engineer, lives in an analog world, where things are continuous. But I, who study information and knowledge, live in a discrete world, where things are divided into relatively small categories and units. This is a fairly fundamental difference and is even reflected in the kind of mathematics that our fields use: His field uses calculus, mine discrete mathematics. Where he uses an integral, I will use a summation. Where he speaks of a differential equation, I will speak of a difference equation. And I make much use of mathematical logic and set theory, tools that tend to be absent from civil engineering, just as complex variables tend to be absent from mine. Neither use of mathematics is better than the other; they are simply suited for different classes of problems.

In a discrete world, small savings do not accumulate. This is usually a virtue, for it means that small errors do not accumulate and become large errors. Library shelves are discrete in that each shelf is relatively short and fixed in length. Saved space on one shelf does no good whatsoever for the next shelf. The real world of artificial objects is more often discrete than continuous, for it is simply much more convenient to have things come in small, convenient packages than in large, continuous ones. And so it goes.

13

Brain Power

MY favorite silly statement, repeated over and over again in many forms, goes like this:

Research now indicates a person probably uses less than one-tenth of 1 percent of his or her brain power.

This particular version of the statement comes from an article in an airline magazine, but different versions of it show up repeatedly, although usually claiming that we use around one-half to one-third.

But what a peculiar statement. What could it possibly mean? How would we even go about measuring such a statistic? Once, out of curiosity, I asked an assembly of eminent cognitive and brain scientists where they thought the statement might have come from, and nobody knew, nor did they have any idea of how they would go about verifying it.

Suppose I say that you use less than half your muscle power. Or that you do not use most of the books in your home. Or TV channels, for that matter. If you have ninety TV channels, you probably watch only about 1 percent of them at a time—in other words only one channel at a time. And since even the most devoted TV family does not watch TV twenty-four hours a day, there is considerable waste. Of course, you might say, it is only possible for a person to watch one channel at a time, or to read one book at a time, or for that matter, to either read or watch TV at any instant of time. So it is silly to say that just because the house has many books or many TV channels with only a few being used at any one time that the "power" of the house is being wasted. Exactly.

What does it mean to say we use only a fraction of the brain? The brain is there and does what it has to do. Why do I care how much of the brain is in use? I only care that it serves me well. The parts of my brain that regulate body temperature, posture, chemical balances, hunger, etc., work however hard they have to work in order to do their job. I wouldn't want them to work any harder.

Of course, the statement is really talking about mental activities—cognitive activities—not the regulation of the heart or the stomach. The assumption here is that much of our brain power for thought is lying fallow, something like unused land on a farm. If only we were less lazy, or better trained, or motivated or whatever, why we could be smarter. Gee, if I use less than one-tenth of 1 percent of my brain, does that mean that if I used it all, I would be a thousand times smarter?

How much of the cognitive part of my brain could I use even if I wanted to? One problem with that percentage is that it doesn't mention time. Does it mean "at any one time" or does it mean "ever?" If it means I use only a tiny fraction of my brain ever, then it is surely false. If it means that at any one time I use only part of my brain, well, it is true. But so what?

Consider the muscles of the body. What percentage of them do I use? Less than half? Probably. After all, most muscles are connected up in an opponent system, so that one muscle opposes or cancels out the other. Normally I would use only one-half of each pair, so normally I am using only half my muscles: and that is exactly how it is supposed to be. And what about my "running muscles"? Are they wasted when I am sitting? Or my lifting muscles? Are they wasted when I am running? Or my—well, you get the point.

If I want to think hard, I have to minimize distractions. I turn off the television set, turn off the radio. When most people are thinking, they stare absentmindedly into space, defocusing the eyes or even closing them so as to avoid seeing anything. This means that the sensory areas of the brain are given little or nothing to do. Notice: The fact that people deliberately reduce extraneous sensory events when trying to think means that, in some sense, to think harder, you deliberately use less of the brain. Does this mean that the less of the brain we use, the smarter we are? No, of course not. The real problem is that attending to events in the world interferes with concentration upon our thoughts.

There has been a large amount of research devoted to these issues. Thus, from studies of attention, we know that when a person tries to do several tasks simultaneously, if they use similar brain structures, the tasks will interfere with one another. A visual task does not interfere with an auditory one. Motor tasks—tasks that require moving of the limbs—do not conflict with visual or auditory ones, unless there are common features. But two auditory tasks will conflict, as will any two tasks that require the processing of information about spatial location, whether they be visual, auditory, or motor. All in all, these studies point to the same conclusion. How much or little of the brain is being used is irrelevant. The critical thing is to get the *correct* parts of the brain involved, without interference or competition from other parts.

Did you know that brain cells die? That although most of the body regenerates new cells to replace dying ones, the brain does not? Once you get into your midtwenties, it is all over: The total number of brain cells starts decreasing.

Does that mean that the older you get, the dumber you become? Some people like to think so, but in general intelligence is not correlated well with age, especially if a person is healthy and active. In fact, those of us who are past thirty prefer to believe we acquire more knowledge as we grow older. There are even scientific studies to back us up (produced, of course, by scientists who are past thirty). So how come we get smarter as we lose brain cells? Mainly because the total number of cells has little to do with intelligence.

The brain is very redundant. Most information is distributed among many cells and locations, although the exact nature of the distribution is not yet fully understood. This distributed nature of knowledge makes the brain very resistant to brain injury. Lose a few thousand cells here and there, and it simply makes no difference.

We have a huge number of brain cells, so many that we can lose millions without ever noticing any difference. Current estimates for the total number of brain cells lie between one hundred billion and one thousand billion. So if we were to lose one million cells, that would be only one-millionth of the total—0.0001 percent. In fact, we could lose one million cells every single day for one hundred years and, at the end, still have 96.4 percent of the original number.

The huge redundancy of the brain is an essential biological survival mechanism. But this means that we must always be using only a fraction of the brain, deliberately so. In any event, I want to assure you that:

> Statements about the percentage of the brain that are in use are meaningless.
>
> Humankind would not suddenly be smarter if it started to use a greater percentage of its brain at any one time.

It doesn't matter how much of your brain is active. What matters is what you are doing with the part that is active.

14

Hofstadter's Law

A FAVORITE question of those writing about technology always is, "How many electric motors do you think you have in your home?" Most people think they may have a couple. Maybe even six or ten. The point of the question, of course, is that electric motors have become an essential part of civilization, so much so that they are hidden away in many devices, leaving us unaware of their presence. As a result, people are always surprised at the real number. Even if you know you're going to have more than you realized, you'll probably have even more than you thought you would, even after taking account of the fact that you'd have more than you realized.

I just went around my home and counted as many as I could discover: eighty-four. Amazing but true, and I probably missed some. Many are in toys, perhaps a dozen or so, but even subtracting these still leaves an impressive number. Clocks and timers account for a large percentage of those motors.

I tried my parents' home, for after all, they live a simpler life and they don't have any toys. Forty-three.

The fact that the number surprises, even after we have been primed to expect a surprise, is a variant of Hofstadter's Law. Doug Hofstadter, a combination physicist, computer scientist, and all around generalist, worried about the fact that estimates on how long it would take to complete a computer program were always short, usually by at least a factor of two. Hofstadter's Law says: "It always takes longer than you expect, even when you take into account Hofstadter's Law." The Law is a brilliant reflection upon human frailty, but it can be extended to cover far richer domains, as the example of electric motors indicates. For the phrase "It always takes longer" we can substitute a description of almost any human endeavor, even apparently contradictory ones.

> ### Hofstadter's Law (revised):
> *It always takes longer*
> *It always costs more*
> *It will always be harder*
> *There will always be more*
> *There will always be less*
> *than you expect,*
> *even when you take into account*
> *Hofstadter's Law.*

Let's test the revised Hofstadter's Law in three new domains: computers, clocks, and batteries. Start with computers. How many computers do you have in your house? The answer depends upon what is meant by "computer," but if any intelligent electronic circuit with program and memory is counted, then the answer is surprisingly large. Someday soon it will exceed even that for electric motors.

Do you count a digital wristwatch with multiple functions? Or the controller for the microwave oven or television or videocassette recorder? How about the smart, programmable thermostat for your home, or even the microprocessor control of sewing machines, washing machines, and driers? Calculators are, of course, small computers. As are fancy telephone instruments, and for that matter, the telephone system itself. The phone system's switching circuits have long been considered by communication engineers as the largest computer system in the world. The television set itself is a computer, and probably the frequency-synthesizer short-wave receiver my family owns would also qualify. The number of computers in the home, car, and office, even today, is large.

In my house, which is clearly unusual in this respect, we have four real computers, plus a laser printer that contains within it a powerful computer, as well as a small local area network to interconnect them. I have more computing power in my home than I had in my advanced, computer-controlled laboratory a mere ten years ago. The twenty-seven clocks and timers in my house add up to a number far larger than I can understand, more than I can use.

But now the ultimate question: "How many *batteries* are there in your house?" I won't be surprised at the number, no matter how large it turns out to be. I'll only be surprised if the number is low. I'm sure the number is in the millions, and not only that, I'm sure that many need replacement.

15

One Chance in a Million

HOW come some people always know just how they got sick? And how come they always expect me to know how I got my cough, or runny nose? Most of the time, there is no way of knowing. The reasons that people tend to give are examples of what we call "folk psychology" or "folk medicine." They have no scientific validity, even though they are a part of popular culture.

Getting wet, or even cold, does not lead to a cold. Yes, we can get cold that way, but this refers to comfort, not to illness. The illness we call a "cold" is not related to how appropriately we dress for the outside temperature. Research with animals has shown that they tend to associate stomach ailments with the last novel food eaten, even if the food was eaten several days prior to the illness. This is probably a good evolutionary rule of thumb—if you feel ill, avoid the last novel food you tried—but it isn't good science. The result is that many of us have learned to dislike perfectly good foods simply because of the accidental correlation of nausea and eating the food for the first time. On the other hand, the opposite rule—the last novel food eaten before feeling better—is thought by some scientists to be responsible for the belief in certain cultures that when ill you should eat chicken soup, or drink mint tea, or any of the other standard folk remedies.

Most colds are minor and run their course in a couple of days. If, however, we are given several big bowls of chicken soup only when ill, then we are apt to associate getting better with eating the soup. The fact that we would have gotten better anyway, or for that matter, that we would have gotten ill in the first place despite our exposure to the cold, or to the damp, or to that funny-tasting food at the party doesn't stop us from trying to find a reason for getting better or falling ill.

When I get sick, I haven't the slightest idea why. I'm just sick, that's all. I don't even care why I am sick. I simply want to get better. But people keep asking: "Where did you get it?" How would I know? I even get asked about other people: "How come George got sick? Do you think it was something at the office?"

I have finally learned how to respond. I now shake my head wisely and say, "There's a lot of that going around lately." What a wonderful statement. On the one hand, it has no meaning whatsoever. On the other hand, people always manage to interpret it so that it is always true. It always satisfies the person to whom I say it—always.

What is "that" anyway, and how can it always be going around? It isn't. But there are always enough cases of colds or flu or other ailments to make the sentence appear meaningful. If I say to you, "There's a lot of that going around lately," you automatically try to think of someone else who has been sick. I guarantee you will almost always recall someone. "Ah yes," you say to yourself, "I just thought of another one." "Oh yeah," you say aloud, "there certainly is."

How likely are you to know someone who is sick? Very likely. A recent newsletter stated that the average American gets a cold three times a year. Suppose we count as "knowing someone who is sick" any time you can think of a person—including yourself—who has been sick in the past week. If you know fifty people, even casually, the chance that just random sickness will cause at least one of them to have been sick within the past week is 95 percent. Those are odds of twenty to one that someone you know will be sick. And in fact, if the circle of people you know or that others will tell you about is larger than fifty, the chance that you will know of someone else who is sick is awfully close to 100 percent. In fact, if you know 100 people, the odds are over 300 to 1 that at least one of them will be sick.*

People are forever underestimating the power of probability. They are surprised when two people they know are sick in the same week and think there is a lot of something going around. But people also make just the opposite mistake: They tend to be very bad at realizing that infrequent events do happen.

Psychologists are fond of showing that people are not very good at estimating the likelihood of rare events. Most people believe airplane travel is far more dangerous than automobile travel even though statistics show the opposite. We see a lot of aircraft accidents reported on television and in the newspapers, but that is precisely because they are so rare. Automobile accidents occur all the time. Most people have been in an accident, or witnessed one. Roughly 40,000 people die each year in the United States from automobile accidents and hundreds of thousands are injured. There are hundreds of accidents every day. As a result, an automobile accident is commonplace, so it isn't likely to be reported in the newspapers.

Aviation, on the other hand, is very safe. This means that when an accident happens, it is important news, so the newspapers and television make a fuss. In 1990 there were only twenty-five accidents with fatal injuries for scheduled, commercial aviation in the entire world. Twenty-five for the entire world. But twenty-five per year is about once every two weeks. Infrequent enough that the accidents make the news, frequent enough that there always seems to be one. Of course, airplane, train, bus, and ship accidents tend to be more costly in terms of lives, money, and ecological impact than the much smaller accidents involving automobiles. Even so, the number of deaths through automobile accidents far exceeds the number of deaths from these other forms of better publicized accidents. Maybe our perception of airplane crashes is something like our perception of colds. All you need is for someone to say, "There certainly seems to be a lot of accidents lately," and if you think about it, why sure enough, you can remember another one. So, "Oh yeah," you say aloud, "yeah, there certainly are."

Airplane crashes are a simple example of relatively rare events that people believe to occur more frequently than they really do. Some events have just the opposite problem: They are thought to be less likely than they are.

Years ago I read an article on the safety of elevators. "Elevators are five times safer than stairs, and the odds of getting stuck in one are 50,000 to 1." Yikes. One chance in fifty thousand? I once worked on the twelfth floor of William James Hall, the home of the psychology department at Harvard University. I took the elevator about ten times a day. Up to my office when I arrived, down again when I left. Down to the library, up to colleagues on a higher floor. Down for lunch. Up back to my office. Down to go teach a class. Back up afterward. Five round-trips a day was not unusual. Even if I had only worked on weekdays that would be about 250 days a year, 2,500 elevator trips a year.

If that reporter was correct, I would be stuck in elevators at least once every twenty years, or two or three times in my lifetime. The interesting thing to me was that the reporter thought the number to be so small that it indicated the great safety of elevators, whereas I, doing some calculation, thought it too high. What the reporter failed to take into account was that even the most unlikely events will happen occasionally if the number of opportunities is high enough.

Nobody is immune to the difficulty of estimating the likelihood of rare events. Even pilots of aircraft make the mistake about their own airplanes. Pilots are apt to be somewhat like the reporter, believing that if something has a very small chance of happening, it is impossible. Perhaps my favorite example comes from an incident where an unlikely sequence of events did happen, but with a happy ending— the plane landed safely. The full story is complex, but worthy of telling. To me, a student of human behavior and aviation safety, it is a wonderful tale of how accidents seldom have a single cause: There is almost always an unlikely sequence of events.

The airplane was a Lockheed L1011, a commercial airliner that looks something like a DC-10 or MD-11. It is about the same size and has three engines, one on each wing and one in the rear of the plane on the tail structure. One morning as it flew its scheduled route from Miami to Nassau in the Bahamas, the instruments reported that one engine was low on oil pressure. The plane was then over the Atlantic Ocean just east of Miami. The pilot dutifully turned off the engine and turned the plane around to return to Miami. As the plane flew along on the remaining two engines, they too began to lose oil pressure. The thought of turning off all three engines while over the ocean understandably upset the flight crew.

The L1011's three engines are not all the same: The two engines on the wings are different from the tail engine. Moreover, they usually are serviced by different people (which was true for this airplane as well). There is no obvious malfunction that could cause all three engines to lose oil pressure at the same time. In fact, the most obvious conclusion would have been that the oil pressure was fine but that the electrical system that powers the indicators was faulty. The pilot therefore said, "It's one chance in a million," and he continued to fly with the other two engines.

The pilot at first decided the problem probably was a faulty electrical system, but that was quickly ruled out by the tests the flight crew performed. So the pilot had no choice but to keep going, hoping to make it back to land. The remaining two engines failed while they were still over water, and the crew prepared for a water landing. However, just in time, they managed to restart the first engine, which they had turned off before it was damaged. It lasted only long enough to land them on the runway, and then failed, leaving them stranded but safe.

The pilot's actions were prudent. Although he could not believe that he had truly lost oil pressure on all three engines, he couldn't have done anything differently even if he had known. The underlying causes of the incident make a classic, textbook case for those interested in the study of human error and how multiple actions and multiple failures lead to unexpected accidents. The L1011 received its usual safety routine: Every evening the plane is checked over by mechanics. In the check, they remove a magnetic plug from the engine and look to see how many metal particles are sticking to the magnet. This is a good way to check on the amount of wear, indicating when the oil should be changed or the engine overhauled. After the plug is removed, there is a hole that allows oil to leak out. This hole, of course, has to be filled with a new plug, and the plug itself sealed with a rubber "O-ring."

Well, on the day of the failures the plane was checked, serviced, and returned to duty. Unfortunately, the mechanics didn't put O-rings on the plugs, so they leaked. All three of them. This despite the fact that two different mechanics serviced the three engines, despite the fact that after servicing, they turned on each engine to check for leaks. Why weren't the O-rings put in place? The mechanics had even signed a checklist claiming they had installed the O-rings.

The problem was, for the many years they had been working at the job, they had never ever installed O-rings. Their supervisor, trying to be helpful, always did it for them. For years, he would get the plugs and the O-rings from the warehouse, put the O-rings on the plugs, and then store the assembled parts in his desk drawer. This made it easier for the mechanics to do their job, so easy in fact, that these mechanics had never needed to get the parts themselves from the warehouse or install the rings on the plugs. The supervisor was doing a useful favor, one that did help the mechanics. Unfortunately, this one night the drawer was empty. The mechanics went to the warehouse to get new plugs, but neglected the O-rings. But why did they sign the checklist? They always did, yet they had never before put on the O-rings.

Who is at fault here? The mechanics? Maybe, for after all, they signed the sheet even though they didn't really do the task. The supervisor? Maybe, for after all, he wasn't supposed to have been so helpful. The warehouse? Maybe, because they had no business handing out the plugs without the O-rings. The system? Yes. A safe system would have made sure that plugs were always packaged with O-rings. It would have permitted the supervisor to help, but the supervisor should then have been the one to sign the checklist, so that the absence of the supervisor's signature should have alerted the mechanics to the fact that something was wrong. The mechanics should not have allowed themselves to have been helped so much: They should have always checked the plugs and rings for themselves.

Was the system at fault for not developing a checking procedure? Yes, the mechanics turned on the engines to look for leaks, but they did not wait long enough. It took a long time for the leak to develop. The people who developed the checking procedure should have actually tested it to make sure it would catch errors of this sort, errors that would have been visible at night, in the dark. (The mechanics had to use flashlights to see what they were doing.)

What about the pilots who couldn't believe that all three engines had failed? Were they at fault? Maybe, although they didn't have too much choice in the matter. Whether they believed or not, they still took the appropriate actions. The real point is that human error strikes in unlikely ways, in unlikely cases. The rule in major accidents is that they occur because of an extremely unlikely combination of extremely unlikely events. However, these unlikely events do happen—and therefore so do accidents.

There is a tendency to believe that unlikely events won't happen. Something that occurs only once every million times is simply not worth worrying about. In some sense, this is true: For the individual person, this is a low risk. But for the nation or the world, one chance in a million is unacceptable. In a country of 250 million people and a world of several billion, one-in-a-million events happen every day.

In the United States alone there are about eight million commercial aviation flights a year. If the chance of a particular accident occurring is one in a million, there will be roughly eight of them a year. One in a million is a very rare event. Eight accidents a year are eight too many.

16

Coffee Cups in the Cockpit

IN 1979 a commuter aircraft crashed while landing at an airport on Cape Cod in Massachusetts.* The captain (the pilot) died and the first officer (the copilot) and six passengers were seriously injured. As the plane was landing, the first officer noted that the plane seemed to be too low, and he told the captain. However, the captain did not respond. The captain, who was also president of the airline and who had just hired the first officer, hardly ever responded. He was the strong, silent type. He was in charge, and that was that. United States airline regulations require pilots to respond to one another, but what was the copilot to do? He was new to the company and the captain was his boss. Moreover, the captain often flew low. There were obvious social pressures upon the first officer.

What the first officer failed to notice was that the captain was "incapacitated." That's technical jargon. What it means is that the captain was unconscious and probably dead from a heart attack. After their investigation of the resulting accident, the U.S. National Transportation Safety Board (NTSB) rather dryly described the incident this way:

> *The first officer testified that he made all the required call-outs except the "no contact" call and that the captain did not acknowledge any of his calls. Because the captain rarely acknowledged calls, even calls such as one dot low (about 50 ft below the 3° glide slope) this lack of response probably would not have alerted the first officer to any physiologic incapacitation of the captain. However, the first officer should have been concerned by the aircraft's steep glidepath, the excessive descent rate, and the high airspeed.*

Seems strange, doesn't it? There they are, flying along, and the captain dies. You'd think the copilot would notice. Nope, it isn't as obvious as you might think. After all, reconsider. During landing, the two pilots are sitting side by side in a noisy airplane with lots of things to do. They hardly ever look at each other. Assuming the captain dies a quiet, polite death, what is there to attract the copilot's attention? Nothing. In fact, United Airlines had tried it out in their simulators. They told the captain to make believe he had died, only to do it very quietly. Then they watched to see how long it took for anyone else to notice. The NTSB reviewed that study in their typically dry fashion:

> *In the United simulator study, when the captain feigned subtle incapacitation while flying the aircraft during an approach, 25 percent of the aircraft hit the "ground." The study also showed a significant reluctance of the first officer to take control of the aircraft. It required between 30 sec and 4 min for the other crewmember to recognize that the captain was incapacitated and to correct the situation.*

So there: The pilot dies and it takes some people as long as four minutes to notice. Even the quick ones took thirty seconds. And a quarter of them crashed (or as the NTSB says, "hit the 'ground'" ["ground" being in quotes because, fortunately, this was a simulator: embarrassment yes, but no injury]).

Commercial aviation is a strange and wondrous place where perceived images are somewhat in conflict with reality. The image is of a heroic, skilled adventurer, successfully navigating a crippled aircraft through storms, fires, and unexpected assaults. Images of Lindbergh and Earhart flying alone over the ocean, or World War I fighter pilots in an open cockpit with helmet, goggles, and scarf still come to mind.

The reality is that the commercial aviation pilot of today is a manager and supervisor, not the daredevil pilot of yore. Today's flight crew must be well schooled in the rules, regulations, and procedures of modern aviation. They are not permitted to deviate from assigned boundaries, and on the whole, if they do their job properly, they will lead a routine and uneventful life. The flight crew is in charge of a large, expensive vehicle carrying hundreds of passengers. The modern flight deck is heavily automated, and multiple color computer screens show maps, instrument readings, and even checklists of the tasks the crew members must do. The flight crew must act as a team, coordinating their actions with each other and with the air traffic control system, in accordance with company and federal policies. Pilots spend much of their time studying the vast array of regulations and procedures. They are tested and observed in the classroom, in the simulator, and in actual flight. Economics and reliability dominate.

As a result of numerous studies of aircraft crews performed by scientists at the National Aeronautics and Space Administration (NASA) and at universities, we have learned a lot about the need for cooperative work and interaction. The lessons actually apply to almost any work situation. It is unwise to rely on an authoritative figure. Individuals can become overloaded and fail to notice critical events. People also have a tendency to focus on a single explanation for events, thereby overlooking other possibilities. It often helps to have several people look things over, encouraging everyone to feel that their contributions are appreciated and encouraged. It isn't a bad idea to consider alternative courses of action. This has to be done with some sensitivity. Suggestions from subordinates have to be accepted as useful and constructive, not as a questioning of authority. This is sometimes hard to pull off, especially where there is a quasi-military authority structure.

The old-fashioned image of the cockpit crew is more like a military hierarchy: captain in charge, first and second officers serving subsidiary roles. Once upon a time it used to be assumed that when an airplane got into trouble, it was the captain's responsibility to fix things. No longer. Another dramatic example of why this philosophy fails comes about from an accident that occurred in 1972 over the Everglades, a swampy, jungle-like region in southern Florida. This particular accident has become famous in the eyes of students of aviation safety and human error. As the plane was coming in for a landing in Miami the crew lowered the lever that lowers the landing gear. However, the lights that indicate a fully lowered and locked gear did not come on. So the crew received permission to circle for a while over the Everglades while they figured out what the problem was.

Now imagine a crowded cockpit with everyone examining the lights, reading through the "abnormal procedures" manual, trying to analyze the situation. Maybe the gear was down and the lights weren't working? How could they tell? Simple, by looking at the gear. Alas, normally the landing gear is not visible from inside the plane, and anyway, it was night. But airplane manufacturers had thought of a solution: You can remove a panel from the floor and lower a little periscope, complete with light, so you can look. Now imagine the entire crew trying to peek. Someone, on hands and knees, is taking off the panel; someone is trying to lower the periscope; someone is reading from the manual. Oops, the light in the periscope breaks. How to fix that? Everyone is so busy solving the various problems that nobody is flying the airplane.

Nobody is flying the airplane? That's not quite as bad as it sounds. It was being flown by the automatic flight controls at constant airspeed and altitude. Except that someone, probably the captain, must have bumped against the control wheel (which looks something like an automobile's steering wheel) while leaning over to watch the other people in the cockpit. Moving the control wheel disconnected the altitude control, the part of the automatic pilot that keeps the airplane at a constant height above ground. When the controls were disconnected, a beep was sounded to notify the crew. You can hear the beep on the tape recording that was automatically made of all sounds in the cockpit. The recorder heard the beep. The accident investigators who listened to the tape heard the beep. Evidently, nobody in the cockpit did.

"Controlled flight into terrain." That's the official jargon for what happens when a plane is flying along with everyone thinking things are perfectly normal, and then "boom," suddenly they have crashed into the ground. No obvious mechanical problem. No diving through the skies, just one second flying along peacefully, the next second dead. That's what happened to the flight over the Everglades: controlled flight into terrain. As nobody watched, the plane slowly, relentlessly, got lower and lower until it flew into the ground. There was a time when this was one of the most common causes of accidents.

Today, I am pleased to report, such cases are rare. Today pilots are taught that when there is trouble, the first thing to do is to fly the plane. I have watched pilots in the simulator distribute the work load effectively. One example I observed provides an excellent demonstration of how you are supposed to do it. In this case, there was a three-person crew flying NASA's Boeing 727 simulator. An electrical generator failed (one of many problems that were about to happen to that flight). The flight engineer (the second officer) noticed the problem and told the captain, who at that moment was flying the airplane. The captain turned to the first officer, explained the situation, reviewed the flight plan, and then turned over the flying task. Then and only then did the captain turn to face the second officer to review the problem and take appropriate action. Newer aircraft no longer have a flight engineer, so there are only two people in the cockpit, but the philosophy of how trouble should be handled is the same: One person is assigned primary responsibility for flying the aircraft, the other primary responsibility for fixing a problem. It is a cooperative effort, but the priorities are set so that someone—usually the captain—is making sure the most important things are being taken care of.

Pilots now are trained to think of themselves as a team of equals. It is everyone's job to review and question what is going on. And it is the captain's job to make sure that everyone takes this seriously. Proper crew resource management probably would have saved the crew and passengers in both the Everglades and the Cape Cod crashes. And it has been credited with several successes in more recent years.

The Strong Silent Type Recurs in Automation

Alas, the lessons about crew resource management have not been fully learned. There is a new breed of strong, silent types now flying our airplanes. For that matter, they are taking over in other industries as well. They are at work on ships, in chemical and nuclear power plants, in factories, and even in the family automobile. Strong silent types that take over controls and then never tell you what is happening until, sometimes, it is too late. In this case, however, I am not referring to people, I am referring to machines.

You would think that the Cape Cod crash and a large bunch of other ones would have taught designers a lesson. Nope. Those crashes were blamed on "human error." After all, it was the pilots who were responsible for the crashes, right? The lesson was not attended to by the folks who designed the mechanical and electronic equipment. They thought it had nothing to do with them.

The real lesson of crew resource management is that it is very important for a group of people doing a task together to communicate effectively and see themselves as a team. As soon as any team member feels superior to the others and takes over control, especially if that person doesn't bother to talk to others on the team and to explain what is happening, then there is apt to be trouble if unexpected events arise.

The Case of the Loss of Engine Power

In 1985 a China Airlines 747 suffered a slow loss of power from its outer right engine. When an engine on a wing goes slower than the others, the plane starts to turn, in this case to the right (technically, this kind of a turn is called "yawing"). But the plane—like most commercial aviation flights—was being controlled by its automatic equipment, in this case, the autopilot, which efficiently compensated for the yaw. The autopilot had no way of knowing that there was a problem with the engine, but it did detect the tendency to turn to the right, which it simply counteracted to keep the plane pointed straight ahead. Negative feedback. (Remember Chapter 6?) Eventually, however, the autopilot reached the limit of how much it could control the yaw and could no longer keep the plane stable. So what did it do? Basically, it gave up.

Imagine the flight crew. Here they were, flying along quietly and peacefully. They had noticed a problem with that right engine, but while they were taking the preliminary steps to identify the problem and cope with it, suddenly the autopilot gave up on them. The plane rolled and went into a vertical dive of 31,500 feet before it could be recovered. That's quite a dive: almost six miles! Ten kilometers! And in a 747. The pilots managed to save the plane, but it was severely damaged. Their safe recovery was much in doubt.

In my opinion, the blame is on the automation: Why was it so quiet? Why couldn't the autopilot indicate that something was wrong? It wouldn't have had to know the cause, it could just say, "Hey folks, this plane keeps wanting to yaw right more than normal." A nice informal, casual comment that something seems to be happening, but it may not be serious. Another question: As the problem persisted, why didn't the autopilot say that it was reaching the end of its limit, thus giving the crew time to prepare. This time perhaps the autopilot should have signaled in a more formal manner because, after all, the problem was getting more serious: "Bing-bong. Captain, sir, I am nearing the limit of my control authority. I will soon be unable to compensate anymore for the increasing tendency to yaw to the right."

The Case of the Fuel Leak

Let me tell you about another case, where again there was trouble developing, but the automatic equipment didn't say anything. This incident was reported in the NASA Aviation Safety Reporting System. These are voluntary reports, filed by people in the aviation community whenever an incident occurs that is potentially a safety problem but doesn't actually turn into an accident (which means there would be no record of the event, otherwise). These are wonderful reports, for they allow safety researchers to get at the precursors to accidents, thereby correcting problems before they occur. In fact, let me tell you about them for a minute.

One accident researcher, James Reason of the University of Manchester, England, calls these reports and other similar signs of possible problems the early symptoms of "resident pathogens." The term comes from analogy to medical conditions. That is, there is some disease-causing agent in the body, a pathogen, and if you can discover it and kill it before it causes the disease, you are far ahead in your aim to keep the patient healthy. "All man-made systems," says Reason, "contain potentially destructive agents, like the pathogens within the human body." According to Reason, they are mostly tolerated, kept in check by various defense mechanisms. But every so often external circumstances arise to combine with the pathogens "in subtle and often unlikely ways to thwart the system's defenses and to bring about its catastrophic breakdown."

This is where those voluntary aviation safety reports come in. They give hints about those pathogens before they turn into full-fledged accidents. If we can prevent accidents by discovering the "resident pathogens" before they cause trouble, we are far ahead in our business of keeping the world a safe place. You would be surprised how difficult this is, however. Most industries would never allow voluntary reports of difficulties or errors by their employees. Why, it might look bad. Will they fix the pathogens that we bring to their attention? "What are you talking about?" I can hear them say, "We've been doing things this way for years. Never caused a problem. That's a million to one chance of a problem." Right, I reply. A million to one isn't good enough.

The aircraft in the incident we are about to discuss is a three-engined commercial plane with a crew of three: captain, first officer (who was the person actually flying at the time the problem was discovered), and second officer (the flight engineer). The airplane had three fuel tanks: tank number 1 in the left wing, tank 2 in the aircraft body, and tank 3 in the right wing. It was very important that fuel be used equally from tanks 1 and 3 so that the aircraft would stay in balance. Here is an excerpt from the aviation safety report:

> *Shortly after level off at 35,000 ft. . . . the second officer brought to my attention that he was feeding fuel to all three engines from the number 2 tank, but was showing a drop in the number 3 tank. I sent the second officer to the cabin to check that side from the window. While he was gone, I noticed that the wheel was cocked to the right and told the first officer who was flying the plane to take the autopilot off and check. When the autopilot was disengaged, the aircraft showed a roll tendency confirming that we actually had an out of balance condition. The second officer returned and said we were losing a large amount of fuel with a swirl pattern of fuel running about mid-wing to the tip, as well as a vapor pattern covering the entire portion of the wing from mid-wing to the fuselage. At this point we were about 2000 lbs. out of balance. . . .*

In this example, the second officer provided the valuable feedback that something seemed wrong with the fuel balance, but not until they were quite far out of balance. The automatic pilot had quietly and efficiently compensated for the resulting weight imbalance. Had the second officer not noticed the fuel discrepancy, the situation would not have been noted until much later, perhaps too late.

The problem was quite serious, by the way. The plane became very difficult to control. The captain reported that "the aircraft was flying so badly at this time that I actually felt that we might lose it." They had to dump fuel overboard from tank 1 to keep the plane balanced and the airplane made an emergency landing at the closest airport. As the captain said at the end of his long, detailed report, "We were very fortunate that all three fuel gauges were functioning and this unbalance condition was caught early. It is likely that extreme out-of-balance would result in loss of the aircraft. On extended over-water flight, you would most certainly end up with . . . a possible ditching."

Why didn't the autopilot signal the crew that it was starting to compensate the balance more than was usual? Technically, this information was available to the crew, because the autopilot flies the airplane by physically moving the real instruments and controls (in this situation, by rotating the control wheel to compensate). In theory, you could see this. In practice, it's not so easy. I know because I tried it.

We replicated this flight in the NASA 727 simulator. I was sitting in the observer's "jump seat," located behind the pilots. The second officer was the NASA scientist who had set up the experiment, so he deliberately did not notify the pilots when he saw the fuel leak develop. I watched the autopilot compensate by turning the wheel more and more to the right. Neither pilot noticed. This was particularly interesting because the first officer had clipped a flight chart to the wheel, so as the wheel tipped to the right, he had to tilt his head so that he could read the chart. But the cockpit was its usual noisy, vibrating self. The flight was simulating turbulence ("light chop"—as described elsewhere in the incident report) so the control wheel kept turning small amounts to the left and right to compensate for the chop. The slow drift of the wheel to the right was visible to me, the observer, but I was looking for it and had nothing else to do. It was too subtle for the pilots who were busy doing the normal flight activities.

Why doesn't the autopilot give more feedback to the crew? Look, suppose that there isn't any autopilot. The first officer is flying, controlling the wheel by hand. The first officer notes that something seems wrong and will probably mention it to the captain. The first officer won't really know what is happening, so it will be a casual comment, perhaps something like this: "Hey, this is peculiar. We seem to be rolling more and more to the left, and I have to keep turning the wheel to the right to compensate." The captain will probably take a look around to see if there is some obvious source of trouble. Maybe it is nothing, just the changing winds. But maybe it is a weight imbalance, maybe a leaky fuel tank. The second officer also hears the comment and so is alerted to scan all the instruments to make sure they are normal. The weight problem would have been caught much earlier.

Communication makes a big difference: You shouldn't have to wait for the full emergency to occur. This is just as true when a machine is working with a person as when two people are working together. When people perform actions, feedback is essential for the appropriate monitoring of those actions, to allow for the detection and correction of errors, and to keep alert. This is hardly a novel point. All automatic controlling equipment has lots of internal feedback to itself. And we all know how important feedback is when we talk to one another. But adequate feedback between machines and people is absent far more than it is present, whether the system be a computer operating system, an autopilot, or a telephone system. In fact, it is rather amazing how such an essential source of information can be left out of system design. Without appropriate feedback, people may not know if their requests have been received, if the actions are being performed properly, or if problems are occurring. Feedback is also essential for learning, both of tasks and also of the way that the system responds to the wide variety of situations it will encounter. The technical jargon for what happens when people are not given proper feedback is that they are "out of the loop."

If the automatic equipment were people, I would say that it was way past time that they learned some social skills. I would advise sending them to class to learn those skills, to learn crew resource management. Obviously it wouldn't do any good to send the equipment, but we certainly ought to send the designers.

Cognitive Science in the Cockpit

The pilots of modern aircraft have lots of different things to do. They have to follow numerous regulations, consult with charts, and communicate with each other, their company, and air traffic control. And then, during the major portion of the flight, they have to sit in the cockpit with very little to do, but keep alert for unexpected problems.

If the trip is long, say across an ocean, there can be many hours of boredom.

Where are the cognitive aids for these tasks? Even the comfort of the flight crew is ignored. Only recently have decent places to hold coffee cups emerged. In older planes the flight engineer has a small desk for writing and for holding manuals, but the pilots don't. In modern planes there are still no places for the pilots to put their charts, their maps, or in some planes, their coffee cups. Where can the crew stretch their legs or do the equivalent of putting their feet up on the desk? And when it is mealtime, how do the crew eat without risk of spilling food and liquids all over the cockpit? The lighting and design of the panels seem like an afterthought, so much so that a standard item of equipment for a flight crew is a flashlight. If physical comfort is ignored, think how badly mental functioning must be treated.

The same is true all over, I might add. I have seen power plants where the operators have to stand on the instruments in order to change burned-out light bulbs, plants where there are no facilities for eating snacks and drinking coffee. It's as if the designers assumed that no equipment would ever burn out, that the workers would go for an entire eight-hour shift without nourishment.

Intuition and anecdote determine the training methods and the development of instrumentation, procedures, and regulations. There has been almost no systematic attempt to provide support for cognitive activities. Crews have to fend for themselves, and because they sometimes recognize their own limitations, many have informally developed routines and "rules of thumb" to help them cope. Consider that many flight crews carry as standard equipment such wonderful cognitive tools as a role of tape and an empty coffee cup to be used as reminders on the instruments and controls. The empty coffee cup is actually quite effective when placed upside down over the throttle or flap handles to remind the pilots that some special condition applies to future use of these controls.

Why Is an Empty Coffee Cup Such a Powerful Cognitive Aid?

One of the most widely studied areas within cognitive science is that of human memory and attention. Although the final scientific theories have yet to be developed, we do know a lot. There is a considerable body of well-understood phenomena and several approximate theories that can be used to good effect in design. Among the simple lessons are the following:

- We can keep only a very limited number of things consciously in mind at any one time in what is called "working memory." How few? Perhaps only five.
- Conscious attention is limited, so much so that it is best to think of a person as being able to focus on only one task at a time. Disruptions of attention, especially those caused by interrupting activities, lead to problems: People forget what was in working memory prior to the interruption; the interruption interferes with performance of the task they were trying to focus on.
- Internal information, "knowledge in the head," is subject to the limits posed by memory and attention. External information, "knowledge in the world," plays important roles in reminding people of the current state of things and of the tasks left to be done. Good design practice, therefore, will provide knowledge in the world.

In general, people are not very accurate at tasks that require great precision and accuracy or precise memorization. People are very good at perceptual tasks, tasks that involve finding similarities (analogies) between one situation and another, and novel or unexpected tasks that involve creative problem solving. Unfortunately, more and more of the tasks in the cockpit force people to do just those tasks they are bad at and detract from the ability to do the things they are so good at.

How do flight crews guard against problems? There are surprisingly few aids. Most of the aids in the cockpit are casual and informal, or invented by crews in response to their own experiences with error. Here are the most common (and effective) cognitive aids in the cockpit:

- Speed bugs: Little metal or plastic tabs that the pilots can move around the outside of the airspeed indicator to help them remember critical settings.
- Crew-provided devices: written notes, coffee cups, and tape.
- Checklists.

SPEED BUGS. Speed bugs are plastic or metal tabs that can be moved over the airspeed indicator to mark critical settings. These are very valuable cognitive aids, for they transform the task performed by the pilot from memorization of critical air speeds to perceptual analysis. The pilots only have to glance at the airspeed and instead of doing a numerical comparison of the airspeed value with a figure in memory, they simply look to see whether the speed indicator is above or below the bug position. The speed bug is an excellent example of a cockpit aid.

The speed bug is an example of something that started out as an informal aid. Some pilots used to carry grease pencils or tape and make marks on the dials. Today the speed bugs are built into the

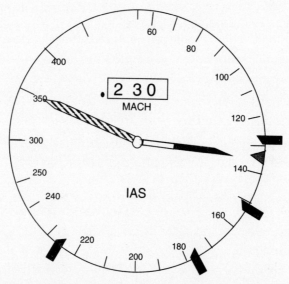

Figure 16.1 An example of a speed bug, an Indicated Air Speed (IAS) instrument for the McDonnell Douglas MD80. The black pointer shows an airspeed of 135 knots (or 0.23 Mach—0.23 the speed of sound at this altitude). The striped indicator at 350 knots indicates maximum permissible speed. The four black marks along the outside are the speed bugs, mechanical pointers that the pilots move to indicate critical speeds for takeoff and landing. The leftward-facing triangle at 132 knots is an internal bug, set automatically.

equipment and setting them is part of standard procedures. Unfortunately, instrument designers have now gotten so carried away by the device that what used to be a single, easy-to-use tool has now been transformed into as many as five or more bugs set all around the dial. As a result, what was once a memory aid has now become a memory burden. I foresee speed bug errors as pilots confuse one bug with another. Newer computer-displayed airspeed indicators sometimes neglect to include speed bugs or other memory aids for critical airspeed settings, sending us back to the dark ages of memory overload.

CREW-PROVIDED DEVICES. Pilots and crews recognize their own memory deficiencies, especially when subject to interruptions. As a result, they use makeshift reminders in the cockpit. In particular, they rely heavily on physical marks. You know, if you want to remember something, tie a string around your finger. Want to remember to take your briefcase? Prop it against the door so you stumble over it when you go out. Want to remember to turn off the air conditioning units before lowering the flaps? Place an empty coffee cup over the flap handle. Crude, but effective.

Pilots need numerous reminders as they fly. They have to remember the flight number, radio frequencies, speed, and altitude clearings. They need to know and remember the weather conditions. They may have special procedures to follow, or special information given to them by air traffic control.

But why hasn't this need been recognized? The need for mental, cognitive assistance should be considered during the design of the cockpit. Why don't we build in devices to help the crew? Instead, we force them to improvise, to tape notes here and there, or even to wedge pieces of paper at the desired locations, all to act as memory aids. The crew needs external information, knowledge in the world, to aid them in their tasks. Surely we can develop better aids than empty coffee cups.

COGNITIVE AIDS. The need for memory aids applies to a wide range of human activities, in all professions and most human activities. Just think about your own activities. How many notes do you write to yourself? Why have Post-it Notes become so popular? How many times have you forgotten something because you weren't reminded at the critical time? In everyday life, these issues are seldom of great importance, but in many industrial settings, they can be critical.

There is a great need for cognitive aids in many aspects of life, aids that are designed with knowledge and understanding of human psychology. Perhaps the one place where these problems are officially recognized is in the checklist, but even this appears to be an anomaly, designed more for the convenience of the training staff than for the needs of the crew.

Checklists

One form of cognitive aid is in widespread existence: the checklist, a list of tasks to be performed. To me, the checklist is an admission of failure, an attempt to correct for people's errors after the fact rather than to design systems that minimize error in the first place.

How are checklists used? In many different ways. In American commercial aviation, the "normal" checklists used in the cockpit are designed solely to serve as checks of actions already done. That is, the cockpit crew first sets up the cockpit for the flight, then uses the checklist to confirm that they did everything properly. Emergencies are treated differently. In these cases, the crew turns to "abnormal checklists" that are read just before or during the required actions. Abnormal checklists serve as reminders guiding and suggesting actions to be performed.

The very fact that checklists exist is admission that not all human behavior is perfect, that errors occur. For safety and thoroughness, some items need to be especially "checked" to ensure that they are done.

Aviation checklists serve multiple functions:

- As checks: to make sure that everything that was supposed to have been done was in fact done. This is actually the primary function of most aviation checklists. This is what the term "check list" really means.
- As "triggers": to remind crews about what needs to be done. This is how the "abnormal checklists" are used in aviation: When problems arise, the crew pulls out the appropriate checklist for the problem and then goes through it, doing each action as it is triggered by the list.

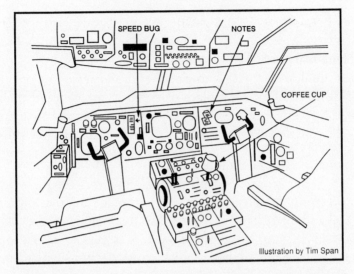

Figure 16.2 Some cognitive aids in the cockpit. An artist's rendering of a hypothetical, modern cockpit, with computer-driven displays replacing the older, mechanical instruments. In this kind of cockpit, airspeed is displayed on a vertically moving "tape," and the speed bug is a pointer alongside the airspeed display. This figure is taken from an earlier article on cognitive aids. It serves as a good illustration, but be aware that, because the artist was not a pilot, the details of the cockpit controls and displays are not accurate.

- For crew communication: In most airlines the entire flight crew takes part in going through the checklist: One person reads the items, one by one, while the others do the specified operations or check the specified state. This procedure ensures that the entire crew knows the state of the aircraft, especially anything unusual. This role of the checklist is often overlooked, but it may be one of its most important functions.
- To satisfy the legal department: Some items on the checklist are really not needed for anything except to guard against lawsuits. After an accident, during the court case, the legal experts want to make sure that the cockpit voice recording will distinctly show the pilots doing certain actions, even if they are not critically needed for flight safety. However, the

longer the checklist, the more chance that there will be
error in using it. Safety demands as short a list as possible;
legal concerns demand longer lists.

But why do we need checklists at all? Not only are checklists a
sign of human fallibility, they are also a sign that the procedures or
equipment design is inappropriate. There are ways to design things to
minimize the chance for skipping critical actions. A properly designed
system might not even require a checklist.

Alas, each new requirement to aid flight crews seems to result in
new tasks for them to do, new procedures to be followed. I can see it
now: My goal of helping the pilots by my analysis of the valuable
memory aid provided by empty coffee cups is misinterpreted so as to
add yet one more item to the procedures and checklists pilots must do.
I can imagine a new checklist item: "Coffee cup supply?" Proper response: "Filled."

Blaming the Person—A Way
to Avoid the Real Issues

One last point: the prevalence to blame incidents on human error.
Human error is the dominant blame for industrial accident. Thus, in
the period 1982 to 1986, the pilot was blamed in 75 percent of fatal
accidents.

Human error. How horrible! What's the matter with those pilots, anyway? Clearly they aren't being trained right. Fire them. Or at
least send them back for more training. Change the training. Add
some more flight regulations. Change the law. Add some more items
to the checklists. This is what I call the "blame and train" philosophy.

Whenever I see such a high percentage of accidents blamed on
individuals, I get very suspicious. When I am told that more than half
of the world's accidents—home and industrial—are blamed on the
people involved, I get very, very suspicious indeed. One way of thinking about the issue is this: If people are only rarely thought to be the
culprit for some problem or accident, then maybe there is some reason
to think that, in the exceptional case, the person did do something
wrong. But if people often seem to be at fault, especially different

people over long periods of time, then the first place to look for the explanation is in the situation itself.

Look, suppose it really is something in people that gives rise to accidents. Shouldn't any sensible designer learn about those things and design the system so it will be resistant to that behavior, or better yet, so it will avoid those situations? Alas, most engineers and designers are not well educated about human psychology. The psychological, cultural, and social knowledge relevant to human behavior is not part of the normal training and education in design or engineering. Moreover, many designers fall prey to the "one chance in a million" syndrome (remember Chapter 15). Until designers take seriously the usability of their designs and realize that inappropriate design is responsible for many accidents and casualties, we will never minimize such incidents. Let me give an example.

In 1988 the Soviet Union's Phobos 1 satellite was lost on its way to Mars. Why? According to the American journal *Science*, "not long after the launch, a ground controller omitted a single letter in a series of digital commands sent to the spacecraft. And by malignant bad luck, that omission caused the code to be mistranslated in such a way as to trigger the test sequence." (The test sequence was stored in the computer memory of the satellite, but it was to be used only during checkout of the spacecraft while on the ground.) Phobos went into a tumble from which it never recovered.

Science wrote its report as if the incompetence of the human controller had caused the problem. The journal interviewed Roald Kremnev, director of the Soviet Union's spacecraft manufacturing plant. Here is how it reported the discussion: "What happened to the controller who made the error? Well, Kremnev told *Science* with a dour expression, he did not go to jail or to Siberia. In fact, it was he who eventually tracked down the error in the code. Nonetheless, said Kremnev, 'he was not able to participate in the later operation of Phobos.'" Both the reporter's question and the answer presuppose the notion of blame. Even though the operator tracked down the error, he was still punished (but at least not exiled). But what about the designers of the language and software or the methods they used? Not mentioned. The problem with this attitude is that it prevents us from learning from the incident, and allows the same problem to be repeated.

This is a typical reaction to a major problem—blame the controller for the error and "malignant bad luck" for the result. Why bad luck—why not bad design? Wasn't the problem the design of the command language that allowed such a simple deviant event to have such serious consequences?

The crazy thing is that normal engineering design practices would never allow such a state of affairs to exist with the nonhuman part of the equipment. All electrical signals are noisy. That is, if the signal is supposed to be 1.00 volt, well, sometimes it will be .94, sometimes 1.18. In a really bad environment—if there are lots of radio transmitters sending energy all about and large electrical cables and motors turning on and off—there might be occasional "spikes" that make the 1-volt signal jump to 5 volts or down to zero or even lower, to -5 volts, for a few thousandths of a second. This noise can really wreck the operation of a computer. Fortunately, there are numerous ways to protect against these problems.

The spacecraft designers—who must design electronic equipment that can function in exactly this kind of a noisy environment—work hard to minimize the effects of noise bursts. If they don't, a burst of noise could turn a signal like 1 volt, which usually would encode the digital code for a "one," into just its opposite meaning, the digital code for a "zero." This could cause untold error in the operation of the computer. But, fortunately, there are many techniques to avoid problems. Designers are expected to use error-detecting and correcting codes.

Suppose, just suppose, that some sort of electrical noise had corrupted the signal sent from the ground to the Phobos satellite, causing it to be destroyed. Who would be at fault then? It certainly wouldn't be the ground controllers. No, the official verdict would probably state that the system designers did not follow standard engineering practice. The next time around that design would be redone to prevent future occurrences. Well, the same lesson applies to all situations: There is no excuse for equipment and procedures that are so sensitive to human error, and certainly no excuse for those so badly designed that they lead humans to err. People err, just as equipment does. Worse, most equipment seems designed so as to lead a person to err:

all those tiny little switches, neatly lined up in an array of identical-looking switches and readouts; computer codes that are long, meaningless strings of letters and digits. Is it any wonder that sometimes the wrong switch gets pushed, the wrong gauge gets read, or a critical character isn't typed? Why would anyone design a computer control language so that a single typing error could lead to catastrophe? And why is it that a procedure meant to be used only on the ground can still be activated once in space? The fault is the design that completely failed to take into account the needs of the people who had to use it. Don't punish the controllers, change the design philosophy.

As automation increasingly takes its place in industry, it is often blamed for causing harm and increasing the chance of human error when failures occur. I propose that the problem is not the presence of automation but rather its inappropriate design. The problem is that the operations under normal operating conditions are performed appropriately, but there is inadequate feedback and interaction with the humans who must control the overall conduct of the task. When unusual situations exceed the capabilities of the automatic equipment, then the inadequate feedback leads to difficulties for the human controllers.

The problem, I suggest, is that the automation is at an intermediate level of intelligence, powerful enough to take over control that used to be done by people but not powerful enough to handle all unusual conditions. Moreover, its level of intelligence is insufficient to provide the continual, appropriate feedback that occurs naturally among human operators. This is the source of the current difficulties. To solve this problem, the automation should either be made less intelligent or more so, but the current level is hazardous. Either the person should be in control all the time or in continual communication with the automatic tools. The intermediate state where the automatic equipment is in control, but with no external feedback, no communication, is probably the worst of all possibilities, especially when trouble strikes.

Complex systems involve a mixture of automatic and human control. Alas, there is too much of a tendency to let the automatic controls do whatever activities they are capable of performing, giving the leftovers to people. This is poor system design. It does not take into account the proper mix of activities, and it completely ignores the needs and talents of people. The price we pay for such disregard for the total system performance comes when things go wrong, when unexpected conditions arise or the machinery breaks down. The total reliability and safety of our systems could be improved if only we treated people with the same respect and dignity that we give to electronic signals and to machines.

17

Writing as Design, Design as Writing

A GRADUATE student of mine, worrying about how to teach the principles of good design to undergraduates, suggested that we should use writing as an example. "We should teach them," he said, "to think of the problem of designing something that people will find understandable and easy to use as the same problem as writing something that other people will understand and find easy to read."

It's a wonderful idea, but it would fail: Most people can't write well. In fact, I often use the idea in the reverse direction: In my attempts to teach students how to write, I use good design as an example. Think of the problem of writing something that other people will understand and find easy to read as the same problem as designing something that people will find understandable and easy to use.

Writing is like design, design is like writing. Although it is useful to try to teach one based upon the properties and needs of the other, either attempt is apt to fail because people tend to be bad at both. To be successful at either task, it is important to be able to take the other person's point of view, to understand that person's background and interests, and to make the work fit the powers and limitations of human cognition. A good designer and a good writer have to share certain characteristics, among the most important being "empathy."

Empathy: understanding, being aware of, being sensitive to,
and vicariously experiencing the feelings, thoughts and experience of another.

But empathy isn't enough. It is easy to be fooled, easy for designers or writers to think they understand their users or readers when in

175

fact, they base their notions solely on their own knowledge and experience. Nothing can be worse than writers or designers who think everyone else is just like them (unless it is designers or writers who think that they are unique and special and nobody else is like them). To be successful, both writing and design have to follow basic psychological principles. And then they must be tested, tried out with readers or users who are similar to the intended audience, and then revised in whatever manner the test results indicate. All this takes a lot of effort and time. Time to learn the principles and appropriate techniques, time to practice them, time to test one's writing or design, time to revise, retest, and re-revise. Few are willing to expend that much time or effort.

Initiation Rites

It is amazing how resistant many people are to the requirements for good design or good writing. The problem is more obvious in writing because so many more people write than design.

One of the things that stands out when talking to long-term users of poorly designed systems is that these people take great pride in their skills. They had to go through great difficulties to master the system, and they are rightfully proud of having done so. That, by itself, is alright. The problem is that the difficulties become a test of the person or group. Then, rather than ease the situation for those who follow, it becomes a sort of initiation rite. The hardy survivors of the experience claim to share a common bond and look with disdain upon those who have not been through the same rites. They share horror stories with one another.

This carries over to much of everyday life. Is the new computer system in the office difficult to learn and to use? Imagine trying to complain to the people who have already mastered the system. "Tough," they will say (or maybe just think to themselves). "It's supposed to be difficult. That's how we separate those with ability from those without. Besides, all of us had to spend a lot of time learning it, losing a lot of work along the way, so why shouldn't you?"

People who use computer systems or complex office machinery (such as copiers and the modern office telephone) will recognize the "initiation rites" syndrome. Those who have mastered the systems feel a great sense of superiority over those who have not. And those who have not, in turn, feel inferior, incompetent, and powerless.

My complaints about computer systems have often encountered this attitude. In the world of computers, there is a software system called "Unix" that many people believe will become popular in the office place. (If you don't know anything about Unix, don't worry, you don't need to. And if you are very lucky, you will never need to.)

Many years ago I wrote an article about the evils of Unix, explaining that it might be a very fine piece of computer science, but as a system intended to be used by ordinary people, it was a disaster. "My secretaries persist only because I insist," I said, pointing out the obscure commands and lack of standards for terminology and procedures, to say nothing of the ways by which Unix could destroy months of work through a moment's mishap. The points I made then are no longer controversial, but at the time, why they were heretical. I was attacked by hundreds of professional programmers across the country. If I didn't approve of Unix, they told me, I had no business using it. Besides, who was I anyway to criticize computer software? In other words, you weren't allowed to criticize unless you were a professional. Being a mere user of the stuff didn't qualify. To my dismay, I had to prove my credentials as a competent scientist and programmer before they would listen further, not that this had anything to do with the merits of Unix.

There is an old English folk saying that goes, "If you can't stand the heat, get out of the kitchen." I have a different approach: Do something about the heat. The folk saying would have us accept the poor designs of the world. Why? After all, if people were responsible for the "heat" in the first place, then people should be able to do something about it. Is the kitchen too hot? Redesign it.

If It's Easy to Understand, Then It Can't Be Very Profound

More than 20 years ago, in an interview, Vonnegut said: "We must acknowledge that the reader is doing something quite difficult for him, and the reason you don't change point of

*view too often is so he won't get lost, and the reason you
paragraph often is so that his eyes won't get tired, so you get
him without him knowing it by making his job easy for him."
I especially love the "get him without him knowing it" part,
but Vonnegut has been almost too successful at that. Among
his more stupid readers are those critics who can't tell the
difference between easy reading and easy writing; because
his books are so easy to read, Vonnegut is accused of "easy"
(or lazy) writing. I think you have to be a writer yourself to
know how hard it is to make something easy to read—or else
you just have to be a little smart.*

Obscurity is often thought of as an essential ingredient in academic writing. Someone who writes clearly is viewed with great suspicion. Basically, the idea appears to be that if the writing is easy to understand, then the ideas beneath it must be inferior: Simple writing reflects a simple mind.

But *These Are Complicated Topics*

One standard excuse of obscurantist authors is that the material in question is complex and technical, sometimes very abstract and refined. The fact that the writing is difficult to follow is unavoidable. The argument then gets turned around: The inability of readers like me to follow such complex thoughts reflects upon me, the reader, not upon the writing. If I really cared, I would do the work required to understand. And if I still can't, well, I should just face up to the fact that my mind isn't sufficiently powerful. Complex ideas require complex writing, and then complex, powerful minds to deal with them. Simple writing is for simple ideas, simple minds.

Is there any case to be made for this? It sounds to me suspiciously like those folks who told me that if I made errors using the Unix computer system, why then I had no business using it. Clearly those who are incompetent to use something or to understand a text have no business trying to do so. Isn't this a great defense? You can cover up any kind of inelegant design or writing this way. Wonderful.

Sure, some thoughts and ideas are complex, but the real test of the power of the idea—and of the thinker—is the ability to translate it into terms that the rest of us can understand. In fact, as long as the ideas and their expression remain convoluted and complex, there is a good chance that they are wrong, even fuzzy-headed. The complexity of the writing masks both the idea and its falsity. In my own case, I once was told that a colleague working on a paper with me had said that "the trouble with Norman's writing is that it's so clear that it's easy to tell when the ideas are wrong."

The implication, of course, is that the correct way to deal with a bad idea is to hide it, disguise it—bury it in indecipherable writing. After all, if an idea is wrong, there are only two things you can do: Fix it or disguise it. The first is very hard, sometimes impossible. The second is easy. (Actually, there is a third thing, but this is hardest of all: Admit that the idea is wrong.)

Now I have to admit that I am treading on dangerous ground here. If all I have is a simple mind, of course I will defend simple writing. After all, if my mind isn't powerful enough to understand the subtleties and complexities of really difficult arguments, then I will want to complain that the difficulty is with the arguments, not with my mind. All my complaints about the difficulty of using so-called "badly designed" things or of understanding so-called "badly written" writing might really be reflections on me, not on the material. Aren't my complaints self-serving?

The issue of deciding who is right—the designer/writer or the user/reader—reminds me of other scientific quarrels. Consider the effect of age on intellectual ability. Young scientists find that human intellectual abilities decline with age, starting in the late twenties or early thirties. However, older scientists doing the same kind of research show that intellectual ability does not decline with age, at least not after you exclude the effects of illness, "cohort" effects, and other technical complexities. Sure, physical and sensory abilities decline—physical strength, vision, hearing, speed of response—but certainly not intellectual ability. In fact, older scientists point out that not only is knowledge unimpaired, it improves with age.

I am similarly reminded of all the studies by middle-aged, white, male professors, administrators, and managers who show that there is no discrimination against people based on age, sex, or color. I am sure you can think of other examples.

Obviously I am not trying to make the point that we are all biased or prejudiced by our ages, professions, social status, or whatever. In fact, that argues against my point: It implies that my argument in favor of clarity of writing probably simply reveals my biases and my own limitations. Nonetheless, I press on with my ways.

Who Should Do the Work, Writer or Reader?

[Hand written] Manuscript culture is producer-oriented, since every individual copy of a work represented great expenditure of an individual copyist's time. Medieval manuscripts are turgid with abbreviations, which favor the copyist although they inconvenience the reader. Print is consumer-oriented, since the individual copies of a work represent a much smaller investment of time: A few hours spent in producing a more readable text will immediately improve thousands upon thousands of copies.

The importance of good writing has changed over time. In the early days of writing, the technology of writing was not easy to master, and there were not very many readers. Imagine how many revisions you would make if the writing was done by chiseling each word into rock. There is no evidence that any serious writing was ever done in quite this fashion, but in the first centuries of writing, the task was not easy. Early writing surfaces were not easy to use: papyrus, leather hides, cloth. Early writing instruments were clumsy and had to be continually tinkered with. Each copy of a book had to be made by hand. No wonder that the author and copyist both did whatever made their task easier, as the opening quotation of this section points out. Who cared about the reader—it was the writer or copyist that mattered. Early books were often designed to look good, as opposed to being readable. Words were broken wherever convenient, the style of typography changed willy-nilly. On the title page of a book, the biggest word might be the first one—even if it was "the." Obviously, its impressive size had no connection with its importance; it just looked better that way.

The first books were designed to simplify the task of the writer or copier, and to make it look pretty to the viewer. The reader was not important. Much reading was done aloud, anyway, so that only one person had to struggle with the text; everyone else had only to listen.

The same points hold today, even though the technology of writing has advanced to the point where the hard work is the mental creation and refinement of thoughts into a form the reader can understand, instead of the physical act of writing. It is easier for writers to let all their thoughts spill out on the page as they materialize than to do the hard, time-consuming work to make those thoughts clear and easy for readers.

Conscientious authors find they must spend considerable time writing and rewriting. An article must be written and rewritten, passed around to students and colleagues, and then revised yet again. It might take months to get a paper in shape for submission to a journal. I once had a friend, a well-respected, young scientist making his way up the academic ranks, who felt he didn't have such time to waste. He had a plan to get promoted, and this required a steady stream of academic papers. He would write a paper once, and that was it. Write it, submit it to the journal, and assuming it got accepted for publication (which it almost always did), rewrite it once more, but only sufficiently to take care of whatever scientific questions or reservations the editor had forwarded. Then he would go on to the next paper (which in fact he had already started even before he had mailed off the previous one).

This scheme worked because scientific journals seldom examine writing style: The editors and reviewers read only for scientific content, looking carefully to see if there are flaws in the arguments, if previous work is properly discussed, if the experiments are well conceived and without obvious artifacts that might have affected the results, and if the interpretation follows from the data. Scientists are not known for their writing ability, and so the editors and reviewers themselves might not even have known why some papers were easier to read than others.

When I complained to my friend that his papers were too difficult to understand and that his talks were packed with too much material for the audience to follow, his response was the classic one of the author-in-a-hurry: "My time is too valuable to spend polishing my papers or talk. It is more important to get on to the next set of ideas. If people want to know what I said or did, everything they need is there: They just may have to work hard, that's all."

The flaw in the argument, of course, is that why should anyone bother? How will they even know that something worthwhile is in there if it cannot be understood? Sure the work got published, and sure, my friend got promoted, but did anyone read the papers? Did his work have any impact? That is where readability matters.

There is an interesting tradeoff between the work required to write well and to read well. The harder the author works, the easier for the reader. Hasty, inconsiderate authors create hours of effort for the reader. Careful, conscientious writers simplify the task for readers, but at the cost of great time and effort for themselves. Whose time is to be worth more: one writer or many readers?

Another way of looking at the issue is to ask how many people are involved on each side of the work. A writer works alone, although actual publication of an article or book will eventually involve a dozen or so people who read the material, approve it, edit it, and often try to patch up the incomprehensible parts. But the potential audience numbers in the thousands, and for really popular works, millions. If one person's work is to be read by that many people, then it is worthwhile to spend the extra time to help them.

Now, in some sense, my academic colleague may have been right about the worth of his efforts. The readership of his articles numbered in the hundreds, and perhaps for such a relatively small number, great effort by him was not required. Then again, maybe his readership was only in the hundreds because of his lack of effort.

Did you know that many scientific articles are probably never read? Whether or not a particular article in a scientific journal is actually read is difficult to tell, but we can determine how many other authors refer to it, because scientific writing requires that one give proper credit to ideas, even ones that argue just the opposite of the point you are trying to make. In fact, opposing ideas are considered very important, because each scientific paper must carefully listen to the opposing voices and try to explain why they are mistaken or why they perhaps do not apply in this particular case. (Or at least, authors have to pretend to pay attention to the opposition, even if privately they scorn and distrust them.)

Because citation of other papers is so important in scientific writing, we can get a sense of how much impact a paper has had by seeing how many other papers refer to it. Most scientific libraries subscribe to the Citation Index, an important reference service that lets you look up a paper and see who else has referred to it. This is a valuable way of learning about research in a scientific specialty: Find an early research paper that is known to be important and then look for other papers that refer to it. This lets you follow the path of further work, up to the current date.

The Citation Index also turns out to be a good way of studying the history, philosophy, and sociology of science itself, allowing the researcher to follow the trail of citations and examine the complex network of interactions and citations. Sometimes scientists look up their own papers in the Citation Index just to see who has referred to them. It is through studies of the Citation Index that we know the interesting statistic that an amazingly large number of scientific papers never receive even a single citation.

But citations are a funny business. Some papers are cited a lot primarily because they are cited a lot. Positive feedback: If a paper is cited once, someone else might thereby hear about it and then refer to it. As more and more people refer to it, the paper becomes known as a standard citation. Others see it mentioned a lot, and then when they want to show that they are scholarly and know what has happened in the past, well, they refer to it also. Even if they themselves have never read it.

Why does this happen? It is common practice in science (and maybe in much of life) to defend one's opinions by reference to authority (see Latour 1987)—as I just did. I made the audacious claim that scientists often back up their claims by referring to some authoritative person who made the same claim rather than by making a logical argument or by demonstrating its validity with experimental evidence. How did I defend that claim? By citing an authority (the French sociologist Bruno Latour).

Latour points out that although every field has a standard set of papers that are agreed to be authoritative, it is all somewhat of a game that scientists play. Just citing the papers is enough: There is no need to have read them. I have discovered that a number of my own articles are sometimes referred to in order to buttress some point the author is making even though my article has nothing to do with the point or, in some cases, argues just the opposite. It's ok though; it's all part of the game. And every so often there is a shift in the way scientists view the phenomena of the field, a new set of authorities appears and, oops, those old authorities are not so authoritative anymore.

For the Benefit of Others

What I find most peculiar about this business of writing and design is that these activities are presumably done for the benefit of others, so shouldn't the needs and abilities of those others be considered? A good writer and a good designer have many things in common. They need to understand the needs and abilities of their audience, and they must consider just how their product will be used.

If you are designing a sophisticated machine for the home, shouldn't you try it out on some homedwellers before manufacture? If you have written an instruction manual for some appliance, shouldn't you ask some prospective appliance owners to use the manual before you print it? Obviously the notion of trying out the material on the intended audience is not popular among designers and writers of instruction manuals, because if it were, we would not have so many unusable products, so many unintelligible instruction manuals.

Now most designers and writers dispute this point. After all, they are fond of pointing out, they are designing things for people, and since they are people themselves, they know just how people work and just what people need. The argument doesn't work. First of all, designers and writers are not ordinary people: They are designers and writers. Designers of kitchen appliances probably spend all day at work designing, not using the kitchen. People who design adding machines or carpentry tools may not be accountants or carpenters. And even if they were, skills at these tasks vary widely, and no single person would be expected to be aware of the wide range of needs of the prospective audience.

Finally, once designers or writers have worked hard on a project, they then know too much about the material to be able to step back and look at it with a neutral eye. Unintelligible sentences seem perfectly reasonable. Even misspelled words or only half-finished sentences can be overlooked. The same is true for design. Writers need to be tested by editors and readers. Designers need this too. Things that are unintelligible or even dangerous to the average user are likely to seem perfectly reasonable to the designer. To find these problems, the evaluation has to be done by people who start with no prior knowledge, no expectations or biases. It is best done by those who are typical of the intended reader or user.

Writing is like design, design is like writing. One serves the reader, the other the user. Both require empathy; both require understanding the needs and abilities of the intended audience. The needs, skills, and desires of readers and users must prevail.

Chapter Notes

CHAPTER NOTES AND BOOK DESIGN

Page
xiv   *The mark "*" appears:* This is an example of a chapter note, in this case containing no useful information.

CHAPTER 1: I GO TO A SIXTH GRADE PLAY

3   *"The best possible way to see an eclipse is simply to LOOK AT IT!":* The description of the frustrated astronomer who missed the solar eclipse while fumbling with his equipment is from Rao's book *Your Guide to the Great Solar Eclipse of 1991* (Rao, 1989, pp. 105–6).

CHAPTER 2: DESIGN FOLLIES

I thank the many people who have aided my search for examples of good and poor design, some of whom have sent photographs, diagrams, and descriptions, others who have let me invade their homes and take photographs and notes.

26   *Figure 2.7:* Alas, I still retain the photograph that was mailed to me from which I made the drawing shown in the chapter, but I have lost the letter. Thank you, (now) anonymous contributor.

27   *Figure 2.8:* Howard Turrentine graciously allowed repeated trips to examine, measure, and photograph his stovetop.

37   *The Canon laser printer* and *[Ford's] foot brake release:* Through electronic mail and the magic of "netnews," Henry Spencer contributed the positive example of Canon's method for avoiding assembly errors; John C. Schultheiss and Paul Raveling contributed the negative examples of Ford's automobile hood and brake release.

Page

CHAPTER 4: REFRIGERATOR DOORS AND MESSAGE CENTERS

49   *In Germany we don't do that:* I thank Gerhard and Ingrid Fischer
      for saving me from my misconception.

49   *A discussion group conducted by electronic mail:* I thank the
      members of the Laboratory for Comparative Human Cognition
      (LCHC) at the University of California, San Diego for originating
      and maintaining the international discussion group, and the "Exter-
      nal" electronic mail discussion group (called XLCHC). I am, of
      course, indebted to the readers of XLCHC for their high-spirited and
      intelligent responses to my question. One reader of XLCHC runs
      yet another international discussion group called the Communica-
      tion Research and Theory Network (CRTNET), and my message got
      forwarded to those readers as well. I got responses from people on
      both these networks, as well as from others who were not members
      of either but whose friends had forwarded my message. All in all, it
      was a fascinating experience.

      Electronic networks are an amazing phenomenon. I first
      learned about the opening of the Berlin Wall from an electronic
      message posted by a jubilant German to XLCHC minutes after it
      happened. The first "issue" of CRTNET I received contained both
      my note asking about refrigerators and an urgent plea from a mem-
      ber of the Serbian Academy of Science for international aid in the
      face of an attack by the Yugoslavian army. (A greater contrast in se-
      riousness of messages could hardly be imagined.)

      One of my Danish correspondents suggested that Jean Lave's
      discussion of everyday cognition was highly relevant to this discus-
      sion: "I think the proper way to describe the communication sociol-
      ogy of the refrigerator door," he said, "would be with the help of
      the concept of 'structuring resources' from Jean Lave's book *Cogni-
      tion in Practice* (1988). The crosscultural discussion of refrigerator
      doors highlights the complexities and concreteness involved in struc-
      turing resources." Lave's book is indeed worthy of examination. It
      is an important book, one I have used in my classes.

50   Excerpts from electronic mail messages: All excerpts have permission
      of the writers to be used in this chapter. Knowing how to treat elec-
      tronic mail messages is a topic for considerable discussion and de-
      bate in the XLCHC community. I promised my respondents that I
      would not quote them or use their names without explicit permis-
      sion. I thank Erik Axel, Dennis Baron, David Barton, Ezio Casari,
      Peter Coughlan, Yrjö Engeström, Rochel Gelman, Russ Hunt, Edwin
      Hutchins, and Michael Schratz.

Page
56   *One British correspondent who has studied communication patterns of everyday people:* This is David Barton, who has published his observations in *Writing in the Community* (Barton and Ivanic, 1991, p. 64).

57   *A program informally called, appropriately enough, "Fridge Door":* Developed by Charles Kerns of Apple Computer, Inc.'s multimedia laboratory in San Francisco. I appreciate the time that was spent describing and demonstrating it.

CHAPTER 5: HIGH-TECHNOLOGY GADGETS

59   *Virtual reality:* A good source is Rheingold's *Virtual Reality* (1991). The only problem is that there are no pictures, so it is hard to imagine what is happening. For pictures, and for a detailed analysis of how these systems will affect art and entertainment, see the excellent book *Artificial Reality* by Myron Krueger (1991), one of the inventors of the technology.

66   *The Japanese Engineer's Design Syndrome:* Lest I be accused of nationalistic chauvinism, I rush to assure you that designers all over the world fall prey to similar or even more devastating design faults and on other occasions I have not hesitated to complain about them. The Americans and Europeans are equally guilty. But the Japanese do seem especially addicted to these tiny little incomprehensible controls and gadgets. I once speculated (in Rheingold, 1991) that perhaps it derives from the Japanese character: People are supposed to go along with the system, not to complain. Therefore the people are supposed to adapt to the technology. My point of view is that technology must adapt to the people.

CHAPTER 6: THE TEDDY

80   *Moravec's Robotic Vision:* Moravec's dreams are described in his book. A more engaging treatment is given by Ed Regis in Chapter 5 ("Postbiological Man") of *Great Mambo Chicken and the Transhuman Condition* (1990). I recommend you start with Regis.

CHAPTER 7: HOW LONG IS NOON?

88   *The engineer Henry Petroski argues:* Petroski's essay on time appears in his book of essays *Beyond Engineering: Essays and Other Attempts to Figure without Equations* (1986).

90   *A curious mix of Latin and English:* From Petroski (1986, p. 143).

Page

CHAPTER 8: REAL TIME

93   Excellent discussions of the history of time can be found in Boorstin's *The Discoverers* (1983); Fraser's *Time, the Familiar Stranger* (1987); and Whitrow's *Time in History: The Evolution of Our General Awareness of Time and Temporal Perspective* (1988).

96   *Baseball tends to be a mysterious game:* Watts and Bahill's (1990) book on baseball, *Keep Your Eye on the Ball: The Science and Folklore of Baseball*, presents an interesting history of how the continual modifications in the rules kept the game at an appropriate level of difficulty as the equipment and the skills of the players changed. (Note especially their table on page 198.)

99   *The direction we call clockwise:* From Feldman's book, *Why Do Clocks Run Clockwise?* (1987).

CHAPTER 10: EVOLUTION VERSUS DESIGN

109   *Is design different from evolution?:* Basalla's book *The Evolution of Technology* (1988) argues that invention proceeds much as does evolution. A number of case histories illustrate how things we think of even today as revolutionary inventions were preceded by a slow series of steps, all leading to the final product.

110   *Bridges, buildings, roadways, and roofs fail:* Petroski provides a fascinating analysis of the lessons learned from structural failures in bridges and other civil engineering structures in his important book *To Engineer Is Human* (1985).

110   *Cargo doors fly open on airplanes:* Airplane failures due to unknown design flaws are rare, but they still happen. The United States National Transportation Safety Board (NTSB) publishes detailed analyses of these incidents. The NTSB reports are noteworthy for their attention to detail and the thoroughness of their investigation. The reports succeed in allowing us to learn from our failures, as they are read and acted upon by people in the aviation industry across the world. For a good example relevant to this chapter, see NTSB (1990a).

CHAPTER 11: TURN SIGNALS ARE THE FACIAL EXPRESSIONS OF AUTOMOBILES

122   *Animal deceit:* The study of deception in animals and children is a booming business among social scientists, for it tells much about the nature of the mind: True deception requires knowledge of another's state of mind. Cheney and Seyfarth's *How Monkeys See the*

Page

World: Inside the Mind of Another Species (1990) provides an excellent treatment of findings in primate studies.

Whiten's *Natural Theories of Mind: Evolution, Development, and Simulation of Everyday Mindreading* (1991) is a good place to begin if you want to know the current state of thought on both animal intelligence and how human infants develop the skills of intelligence, social interaction, and the understanding that the beliefs and knowledge of others may differ from their own.

123–4   *A male vervet, Kitui, gave leopard alarms:* The two quotations come from Jolly's (1991, p. 574) review of the book by Cheney and Seyfarth. Interested readers should see the book itself. In addition, an article by Premack and Woodruff (1978) provides an excellent introduction to the complex manner of studying the self-knowledge of animals.

130   *In Mexico, one wins by aggression:* Please note that I do not wish to imply that the driving habits of cultures carry over to other behaviors. Mexicans are certainly not more aggressive a people than the British. If anything, they have been much more the victims of aggression. But the differences in driving behavior are striking.

CHAPTER 12: BOOK JACKETS AND SCIENCE

135   *He measured the thickness of book jackets:* The chapter was inspired by the musings of Petroski in *Beyond Engineering: Essays and Other Attempts to Figure without Equations* (1986). To be fair to Petroski, his discussion of the space taken by book jackets was just a minor comment in his essay on the history, function, and artistic value of jackets and the book covers that lie hidden beneath.

CHAPTER 13: BRAIN POWER

140   *Research now indicates:* From an article by Grossman (1988, p. 74).

CHAPTER 14: HOFSTADTER'S LAW

144   *Hofstadter's Law:* From Hofstadter's *Gödel, Escher, Bach: An Eternal Golden Braid* (1979, p. 152).

CHAPTER 15: ONE CHANCE IN A MILLION

148   *The average American gets a cold three times a year:* The number comes from an article in the newsletter *Bottom Line Personal* for March 30, 1991, attributed to David Fairbanks, M.D., quoted in *Working Mother* (no date given). A "fact" from a relatively un-

Page

known newsletter quoting an unknown physician who is reported to have said something in an unknown publication is not particularly reliable, but the number itself seems reasonable.

148   *If you know 100 people, the odds are over 300 to 1 that at least one of them will be sick:* If the three times of the fifty-two weeks in the year the average person is sick are randomly distributed across the year, then the chance of being sick for any particular week is 3/52. The chance of being well for any particular week is, therefore, 49/52. If you know 50 people, the chance that all 50 have been well is just 49/52 multiplied by itself 50 times, or $(49/52)^{50} = 0.05$, which is 5 chances in 100. That's only a 5 percent chance that all are well, or if you like, a 95 percent chance that at least one is ill. The odds are 20 to 1 that one of your fifty acquaintances is ill this week.

     If you know 100 people, the chance that all of them are well this week is $(49/52)^{100}$, or only 0.003. This means the chance that at least one of them is sick this week is 99.7 percent: odds of over 300 to 1.

149   *The odds of getting stuck in one are 50,000 to 1:* Nelson (1983).

150   *The instruments reported that one engine was low on oil pressure:* The aviation incident is described in detail in the National Transportation Safety Board's report of the incident (NTSB 1984).

     An excellent review of the scientific literature on these topics is provided by Gilovich, *How We Know What Isn't So: The Fallibility of Human Reason in Everyday Life* (1991).

CHAPTER 16: COFFEE CUPS IN THE COCKPIT

154   Please do not think that because most of my examples are from aviation that air travel is unsafe or that the same thing doesn't happen elsewhere. Aviation is very safe because of the careful investigations of all accidents. The causes are examined with great care and thoroughness and the conclusions are taken very seriously by the aviation community. The voluntary aviation safety reporting system is a major source of safety lessons. It is essential that this service be maintained and I recommend it to other industries.

     Since other industrial areas don't receive the same care and analysis as does aviation, I can't turn to them as readily for examples. But the problems are there, I assure you. Moreover, the problems are often worse. Just look at all those railroad accidents, industrial manufacturing plant explosions, toxic chemical leaks, nu-

Page

clear power incidents, ship leaks and collisions. Aviation is extremely safe compared to these other areas. I refer you to Perrow's *Normal Accidents* (1984), and an excellent review of aviation issues in Wiener and Nagel's *Human Factors in Aviation* (1988).

154   *A commuter aircraft crashed while landing at an airport on Cape Cod:* From NTSB (1980).

156   *An accident that occurred in 1972 over the Everglades:* NTSB (1973). It turned out that the landing gear had been lowered properly, but the light bulb that would have informed the cockpit crew was burned out.

158   *Pilots now are trained to think of themselves as a team of equals:* See Foushee and Helmreich *Group Interaction and Flight Crew Performance* (1988).

158   *The Strong Silent Type Recurs in Automation:* The social issues that affect how automation interacts with workers and changes the nature of jobs are discussed in Zuboff's *In the Age of the Smart Machine: The Future of Work and Power* (1988). Her distinction between "automating" and "informating" is particularly relevant.

159   *In 1985 a China Airlines 747 suffered a slow loss of power from its outer right engine:* NTSB (1986) Also see Wiener (1988).

160   *Early symptoms of "resident pathogens":* See Reason's *Human Error* (1990). A symposium on these issues for high-risk industries was presented in 1990 at the Royal Society, London, and published as Broadbent, Baddeley, and Reason, *Human Factors in Hazardous Situations* (1990). It contains a very early version of this essay.

160   *"All man-made systems," says Reason:* Reason (1990, p. 197).

161   *Shortly after level off at 35,000 ft.:* The voluntary reporting incident was "Data Report 64441, dated Feb. 1987." (The records are anonymous, so except for the date and the fact that this was a "large" commercial aircraft, no other information is available. This is proper, in my opinion, for such anonymity is essential to maintain the cooperation of the aviation community.)

162   *We replicated this flight in the NASA 727 simulator:* The simulated flights that I refer to were done at the Boeing 727 simulator facilities at NASA-Ames in studies conducted by Everett Palmer of NASA, who also serves as the "project manager" for our research grant.

166   *An example of a speed bug:* Figure is redrawn from Tenney (1988).

168   *Checklists:* My analyses of the checklists were done jointly with my colleague, Edwin Hutchins. Some of the information and analyses come from the excellent review by Degani and Wiener (1990).

Page

169 *Some cognitive aids in the cockpit:* From Norman (1991), with permission.

170 *The pilot was blamed in 75 percent of fatal accidents:* From *Annual Review of Aircraft Accident Data. U.S. Air Carrier Operations Calendar Year 1987.* (NTSB 1990b).

171 *In 1988 the Soviet Union's Phobos 1 satellite was lost on its way to Mars:* The report came from an editorial in *Science* magazine (Waldrop 1989), and it formed the basis of an opinion piece by me (Norman 1990). The article appeared in the *Communications of the ACM*, the primary journal for American computer scientists. (ACM stands for Association for Computing Machinery.) I argued that computer scientists need to pay a lot more attention to the design of their programs, or else they too will cause Phobos-like accidents.

CHAPTER 17: WRITING AS DESIGN, DESIGN AS WRITING

177–8 *More than 20 years ago, in an interview, Vonnegut said:* From a commentary by American novelist John Irving on fellow novelist Kurt Vonnegut. (Irving 1990).

180 *[Hand written] Manuscript culture is producer-oriented:* From Ong's *Orality and Literacy* (1982, pp. 122–3).

183 *(see Latour 1987):* Latour's book, *Science in Action*, really is an authoritative analysis of the way scientists work. I cite it here as an illustration as well as to suggest that you might find it interesting.

References

Barton, D., and Ivanic, R. (Eds.). (1991). *Writing in the community*. London: Sage.

Basalla, G. (1988). *The evolution of technology*. New York: Cambridge University Press.

Boorstin, D. J. (1983). *The discoverers*. New York: Random House.

Broadbent, D. E., Baddeley, A., and Reason, J. T. (Eds.). (1990). *Human factors in hazardous situations*. Oxford: Oxford University Press.

Cheney, D. L., and Seyfarth, R. M. (1990). *How monkeys see the world: Inside the mind of another species*. Chicago: University of Chicago Press.

Degani, A., and Wiener, E. L. (1990). *Human factors of flight-deck checklists: The normal checklist* (NASA Contractor Report 177549). Moffett Field, CA: National Aeronautics and Space Administration, Ames Research Center.

Feldman, D. (1987). *Why do clocks run clockwise? And other imponderables*. New York: Harper & Row.

Foushee, H. C., and Helmreich, R. L. (1988). Group interaction and flight crew performance. In E. L. Wiener and D. C. Nagel (Eds.), *Human factors in aviation*. Orlando, FL: Academic Press.

Fraser, J. T. (1987). *Time, the familiar stranger*. Redmond, WA: Tempus Books.

Gilovich, T. (1991). *How we know what isn't so: The fallibility of human reason in everyday life*. New York: Free Press.

Grossman, J. (1988, September). Mind mapping. *USAIR In-Flight Magazine*, p. 74.

Hofstadter, D. R. (1979). *Gödel, Escher, Bach: An eternal golden braid*. New York: Basic Books.

Irving, J. (1990, September 2). Vonnegut in prison and awaiting trial [Review of *Hocus Pocus* by Kurt Vonnegut]. *Los Angeles Times Book Review*, p. 1.

Jolly, A. (1991). Thinking like a vervet. [Review of Cheney & Seyfarth (1990), *How monkeys see the world: Inside the mind of another species*]. *Science*, 251, 574–575.

Krueger, M. W. (1991). *Artificial reality* (2nd ed.). Reading, MA: Addison-Wesley.

Latour, B. (1987). *Science in action*. Cambridge, MA: Harvard University Press.

Lave, J. (1988). *Cognition in practice*. Cambridge, UK: Cambridge University Press.

Moravec, H. (1988). *Mind children: The future of robot and human intelligence*. Cambridge, MA: Harvard University Press.

NTSB. (1973). *Aircraft accident report—Eastern Air Lines, Inc., Lockheed L-1011, Miami International Airport, Miami, Florida, December 29, 1972* (Report No. NTSB/AAR-73/14). Washington, DC.

NTSB. (1980). *Aircraft accident report—Air New England, Inc., DeHavilland DHC-6-300, N383 EX Hyannis, Massachusetts, June 17, 1979* (Report No. NTSB/AAR-80/01). Washington, DC.

NTSB. (1984). *Aircraft accident report—Eastern Air Lines, Inc., Lockheed L-1011, N334EA, Miami International Airport, Miami, Florida, May 5, 1983* (Report No. NTSB/AAR-84/04). Washington, DC.

NTSB. (1986). *Aircraft accident report—China Airlines 747-SP, N4522V, 300 nautical miles northwest of San Francisco, California, February 19, 1985* (Report No. NTSB/AAR-86/03). Washington, DC.

NTSB. (1990a). *Aircraft accident report—United Air Lines Flight 811, Boeing 747-122, N4713U, Honolulu, Hawaii, February 24, 1989* (Report No. NTSB/AAR-90/01, Govt. Accession No. PB 90/910401). Washington, DC.

NTSB. (1990b). *Annual review of aircraft accident data. U.S. air carrier operations calendar year 1987* (Report No. NTSB/ARC-90/01, Govt. Accession No. PB 91/119693). Washington, DC.

Nelson, P. (1983, September). Upward mobility. *TWA Ambassador*, p. 49.

Norman, D. A. (1990a). Commentary: Human error and the design of computer systems. *Communications of the ACM, 33*, 4–7.

Norman, D. A. (1990b). *The design of everyday things*. New York: Doubleday. [Originally published as Norman, D. A. (1988). *The psychology of everyday things*. New York: Basic Books.]

Norman, D. A. (1991). Cognitive science in the cockpit. *CSERIAC Gateway, 11*(2), 1–6. (A publication of the Crew System Ergonomics Information Analysis Center, managed by the University of Dayton Research Institute, Dayton, Ohio.)

Ong, W. J. (1982). *Orality and literacy: The technologizing of the world*. London: Methuen.

Perrow, C. (1984). *Normal accidents*. New York: Basic Books.

Petroski, H. (1985). *To engineer is human: The role of failure in successful design*. New York: St. Martin's Press.

Petroski, H. (1986). *Beyond engineering: Essays and other attempts to figure without equations*. New York: St. Martin's Press.

Premack, D., and Dasser, V. (1991). Perceptual origins and conceptual evidence for theory of mind in apes and children. In A. Whiten (Ed.), *Natural theories of mind: Evolution, development and simulation of everyday mindreading*. Cambridge, MA: Blackwell.

References

Premack, D., and Woodruff, G. (1978). Does the chimpanzee have a theory of mind? *Behavioral and Brain Sciences, 1,* 515–526.

Rao, J. (1989). *Your guide to the great solar eclipse of 1991.* Cambridge, MA: Sky Publishing.

Reason, J. (1990). *Human error.* Cambridge, UK: Cambridge University Press.

Regis, E. (1990). *Great mambo chicken and the transhuman condition.* Reading, MA: Addison-Wesley.

Rheingold, H. (1991). *Virtual reality.* New York: Summit Books.

Tenney, D. (1988, December). Bug speeds pinpointed by autothrottles mean less jockeying but more thinking. *Professional Pilot,* pp. 96–99.

Waldrop, M. M. (1989). Phobos at Mars: A dramatic view—and then failure. *Science, 245,* 1044–1045.

Watts, R. G., and Bahill, A. T. (1990). *Keep your eye on the ball: The science and folklore of baseball.* New York: W. H. Freeman.

Whiten, A. (Ed.). (1991). *Natural theories of mind: Evolution, development and simulation of everyday mindreading.* Cambridge, MA: Blackwell.

Whitrow, G. J. (1988). *Time in history: The evolution of our general awareness of time and temporal perspective.* Oxford: Oxford University Press.

Wiener, E. L. (1988). Cockpit automation. In E. L. Wiener and D. C. Nagel (Eds.), *Human factors in aviation.* Orlando, FL: Academic Press.

Wiener, E. L., and Nagel, D. C. (Eds.). (1988). *Human factors in aviation.* Orlando, FL: Academic Press.

Zuboff, S. (1988). *In the age of the smart machine: The future of work and power.* New York: Basic Books.

Index